Global Production Networks

Operations Design and Management

Second Edition

Global Production Networks

Operations Design and Management

Second Edition

Edited by Ander Errasti

CRC Press
Taylor & Francis Group
Boca Raton London New York

CRC Press is an imprint of the
Taylor & Francis Group, an **informa** business

CRC Press
Taylor & Francis Group
6000 Broken Sound Parkway NW, Suite 300
Boca Raton, FL 33487-2742

First issued in paperback 2017

© 2013 by Taylor & Francis Group, LLC
CRC Press is an imprint of Taylor & Francis Group, an Informa business

No claim to original U.S. Government works

ISBN-13: 978-1-4665-6292-9 (hbk)
ISBN-13: 978-1-138-07727-0 (pbk)

Library of Congress Cataloging-in-Publication Data

Errasti, Ander.
    Global production networks : operations design and management / Ander Errasti.
-- 2nd ed.
        p. cm.
    Includes bibliographical references and index.
    ISBN 978-1-4665-6292-9
    1. International business enterprises--Management. 2. Business logistics. 3. Globalization. I. Title.

HD62.4.E773 2013
658.5--dc23   2012029705

Visit the Taylor & Francis Web site at
http://www.taylorandfrancis.com

and the CRC Press Web site at
http://www.crcpress.com

Contents

Preface

This book, *Global Production Networks: Operations Design and Management,* intends to answer new questions and challenges that companies face when designing and implementing global manufacturing and logistic networks. In a more and more dynamic and volatile market environment, production networks need to evolve over time and local approaches have to be integrated and coordinated with global infrastructures.

The phenomenon of globalization has increased in the past decades due to the opening of borders in Eastern Europe and the sudden emergence of other countries in the global trade economy. The companies that started the internationalization process in the past 10 years have designed value chains that cross different countries in international multisite manufacturing networks with complex management systems to gain efficiency and effectiveness.

The internationalization process is one of the most complex decisions that SMEs (small and medium enterprises) and Business Units from industrial divisions undertake, and it requires companies to balance opportunities and threats when designing and managing an operations strategy. The need to make more effective decisions is more crucial for SMEs and Business Units from industrial divisions because their economic resources and the possibility to acquire capabilities to accomplish these new challenges are limited. Unlike multinational enterprises, SMEs are less familiar with the need to coordinate and integrate activities fragmented in different parts of the world.

The **GlobOpe** (**Glob**al **Ope**rations) framework and roadmap summarizes the contribution of different authors and experiences in Global Operations. This enables managers and practitioners to start the operation's internationalization process in an SME or Business Unit and increase their chances for success.

What Are the Book's Aims?

- To collect the best contributions from researchers in the area of Global Operations and Internationalization Process regarding the principles, tools, and techniques that help managers and practitioners to tackle the design and management of a Global Manufacturing and Logistics Network.

- To present case studies that show best practices and current trends. This enables companies to assess and balance the incorporation of these practices into their operations strategy deployment roadmap.
- To summarize all the contributions to this volume into a framework called GlobOpe, which serves as a useful guideline for industrial SMEs and Business Units.

Who Is This Book for?

This book is geared toward managers and chief operations officers who need guidelines for analyzing, assessing, defining, and deploying their company's operations strategy.

The contributions in this volume also will be useful for consultants and researchers who need to deepen their knowledge of the most innovative tools and techniques.

Finally, lecturers and MBA students who need to know the state-of-the-art and case studies in order to gain better understanding of and insights into issues in Global Operations will find this book invaluable.

Acknowledgments

My research and writing in Operations Management spans a number of years, and it is a pleasure to thank all the colleagues, friends, family, and mentors who have helped me in so many ways to produce this book as a result of this effort.

My first thoughts are for my parents, Jose Luis and Mirentxu, for educating me in the right values with love and patience. Also for my sisters, cousins, nephews, and, especially, Sonia, Jon, and Maite for their support and amazing family life that makes sense of the efforts and sacrifices.

I also would like to thank FAGOR Household Appliances, Mondragon School of Engineering, ULMA Handling, TECNUN School of Engineering, and NATRA Group for supporting me in my industrial and academic career.

In the industry, I have met incredible and outstanding managers from different companies where I have worked on consultancy assignments, such as ULMA Group, URSSA, EROSKI, ORONA, Lanik, Bosch Siemens, Indar Ingeteam, Corrugados Azpeitia, CIE Recyde, UVESCO, ASTORE, TERNUA, and NATRA group. Thank you very much for allowing me to contribute as an advisor.

I am also very grateful to the contributors of this book, especially the researchers in Global Operations from TECNUN and the managers from relevant companies, such as IRIZAR, DANOBAT, CAF, etc., who have aided in enriching the contents with their expertise.

Last, but not least, I would like to thank as well the language service of TECNUN and O'CLOCK language school for the support and commitment to this project.

I hope that this contribution will help managers accomplish their new challenges in the Global Operations context.

Non gogoa han zangoa.
(Basque proverb and TERNUA's community vision meaning:
Where your thoughts go, your steps will follow.)

Ander Errasti

Introduction

The current situation of the world economy due to the last whiplashes of the recession still being very obvious in some countries, the BRIC (Brazil, Russia, India, and China) and other emergent countries expanding their influence, and the international finance system being rethought, obliges us to reflect on which strategies are most suitable to successfully tackling the new challenges that are facing us.

Faced with this new panorama, companies need tools to help them make decisions that will bring greater guarantees and avoid risks. *Global Production Networks: Operations Design and Management*, thus, aims to facilitate decision making for companies in the design and configuration of their global operations overseas. The book, concerned with the internationalization of operations, responds to new questions and challenges that industrial companies have when faced with acting in global markets.

The process of internationalization in global markets context has often been tackled from the business point of view, but rarely from the perspective of the production and logistics systems that support it. In this book, we look in-depth at the strategy of production and logistics operations and endeavor to be a guide for those managers who need to analyze, assess, define, and deploy the operations strategy in their companies.

Trends in the internationalization of operations and reasons for starting the process are presented. Also, factors that determine the location of new production units are studied, as well as methodological aspects to analyze whether to carry out or subcontract part of the production process and the local or global configuration of the supply chain.

Also included are the possible roles or strategic functions that a production plant can take on and the characteristics of the value chain that is developed to reach these functions. Furthermore, the methods used in the physical design of a plant are looked into, together with aspects that need to be considered due to the specificity of a new production plant with regard to the mother company.

In addition, the book deals with how to design and develop the strategic purchasing function and associated local–global supplier network, thus presenting tools that make it possible to assess purchasing policies and establish levers to improve the integral cost of supplies.

The book also looks at a basic area, such as people management, which is fundamental for the success of projects abroad.

The entire book is synthesized in the **GlobOpe (Global Ope**rations) Model, which, without doubt, is going to be very useful for companies immersed in the internationalization processes.

The Editor

Ander Errasti, PhD, is senior lecturer and researcher of supply chain and operations management at TECNUN, the School of Engineering at the University of Navarra, San Sebastian, Spain. He worked in the Engineering Purchasing Department at FAGOR Household Appliances and at the University of Mondragon in Spain. He received his PhD in SME Supply Chain Strategy Diagnoses and Deployment at TECNUN and then worked as business  manager in the Consultancy Business Unit of ULMA Handling Systems. During this period, Errasti worked as senior consultant with account management responsibility with major companies (ULMA Group, ARCELOR Corrugados, Cie Recyde Group, EROSKI Group, Maier, URSSA, Lanik, etc.) from different sectors, such as distribution, automotive, and equipment goods. As a lecturer, he gives lectures in Operations Management in MBA and MBA-executive programs. He also is actively involved in a number of research projects and consultancy assignments that directly relate to his research. Errasti has published several papers in journals, international conferences, and for seminars oriented toward practitioners. He has recently published an outstanding book in Spanish entitled *Warehouse Design and Management* and *Purchasing Management*. He works as Operations Deputy Officer for NATRA Group.

Contributors

Tim Baines

Baines is the professor of Operations Strategy at Aston Business School in England. He received an MSc in Manufacturing Systems Engineering and PhD in Manufacturing Strategy Formulation, both from Cranfield University. He specializes in the realization of competitive manufacturing operations. He is highly active in postgraduate and executive teaching, has experience in a wide range of industrial engineering, and works with the leading companies  in his field, including Rolls-Royce, Caterpillar, Alstom, MAN, and Xerox. His career started with a technician apprenticeship, and has progressed through a variety of industrial and academic positions, including that of Visiting Scholar within the Center for Technology, Policy, and Industrial Development at the Massachusetts Institute of Technology. Baines holds a variety of positions, including member of the EPSRC College of Peers, Fellow of both the Institution of Mechanical Engineers and the Institution of Engineers and Technologists, and a Chartered Engineer.

Claudia Chackelson

Chackelson is an industrial engineer and joined the Industrial Management Department of TECNUN, University of Navarra as a PhD candidate in 2009. To date, her research is focused on World Class Warehousing. Previously, she participated in research projects related to transport, cost, and forestry at the Research and Technological Innovation Center of Montevideo, University of Montevideo (Uruguay).

Donatella Corti

Corti received a master's degree with honors in Production and Management Engineering from the Politecnico di Milano (Italy) and received a PhD in the Department of Economics, Management, and Industrial Engineering at the same university. Since 2005, Corti has been a researcher and assistant professor in the same department. A teacher of Production Systems Management, Operations Management, and Quality Management, her main research interests are related to the globalization of operations, the servitization of manufacturing companies, and mass customization. She has participated in national and international funded research projects and has authored several articles published in international journals and conference proceedings.

Migel Mari Egaña

Egaña received his degree in Industrial Management Engineering from the Mondragon Polytechnic School at the University of Mondragon (Spain). He coordinates several MBA programs and has actively worked on different European and regional research projects on manufacturing systems optimization. Egaña worked as a senior consultant with industrial companies, such as ARCELOR and CEGASA, improving performance through Six Sigma and Lean Production principles and techniques.

Jose Alberto Eguren

Eguren is an industrial organization engineer from Mondragon Polytechnic School/University of Mondragon, Spain. He is a senior lecturer at Mondragon and his research area is related to quality management and continuous improvement. Eguren worked as a quality manager at Torunsa (Spain) and he also has worked on consultancy assignments with several components manufacturers in the automotive and household appliance sectors, such as ORKLI and FAGOR EDERLAN. He received his PhD in a framework for developing sustainable continuous improvement programs in SMEs.

Carmen Jaca

Jaca is a lecturer at TECNUN (University of Navarra, Spain). She studied Industrial Engineering at the University of Navarra, and her research activities are in Continuous Improvement and Teamwork. Jaca also received her PhD in Industrial Engineering from TECNUN. Prior to coming to TECNUN, she worked in a number of industrial companies as a quality manager. She is a member of the board of the QMOD-ICQSS International Conference on Quality Committee.

Bart Kamp

Kamp received a degree in Sciences Politics from Radboud University and a degree in Management in Tilburg University (both in The Netherlands). He has led several research projects related to business and regions competitiveness. Kamp is currently head of the Strategy Department at the Basque Institute of Competitiveness/ORKESTRA (Spain).

Sandra Martínez

Martínez received her undergraduate degree in Industrial Management Engineering from TECNUN, the School of Engineering at the University of Navarra, in 2009. Currently, she is working toward her PhD in the Industrial Organization Department at TECNUN. She belongs to the International Production and Logistic Networks and Logistics Operations Research Group. Her research focus is on operations strategy analysis aided by simulation techniques, the ramp-up processes for new facilities, and the design and configuration process of a global production and logistic network.

Miguel Mediavilla

Mediavilla holds an undergraduate degree in Industrial Engineering from Mondragon Polytechnic School (Spain) and is currently completing his PhD studies in International Operations Management. Mediavilla spent his professional career at BSH Bosch and Siemens Home Appliances Group, where he worked in different positions in supplier quality, industrial engineering, Six Sigma, and production, both in Spain and Germany, before reaching his current position of senior project manager for productivity improvement and Lean management, working in Europe, Asia, and the Americas.

Kepa Mendibil

Mendibil is a lecturer at the Strathclyde Institute of Operations Management, University of Strathclyde (Scotland). He has led a number of international and national research and consultancy programs, and worked for industrial organizations on business improvement, process re-engineering, and ERP (enterprise resource planning) implementation projects. Mendibil's research interests include high value manufacturing and innovation systems, and his work has been published in academic journals and conference proceedings.

Torbjørn Netland

Netland is a PhD candidate at the Norwegian University of Science and Technology (NTNU) in Trondheim, Norway. He was granted a Fulbright scholarship as a visiting researcher at Georgetown University, Washington D.C., for the 2011–2012 academic year. His field of expertise is the use of global improvement programs and company-specific production systems in large manufacturing companies. Prior to his PhD studies, Netland worked for many years as a research scientist and project manager at the SINTEF Research Institute (Trondheim, Norway), where he conducted research with international companies, such as Volvo Aero, Kongsberg Defence & Aerospace, Norsk Hydro, and Pipelife.

Raul Poler

Poler is a professor of Operations Management and Operations Research at the Universidad Politécnica of Valencia, Spain. He is deputy director of the Research Centre on Production Management and Engineering (CIGIP). Poler has led several European research projects and published a hundred research papers in a number of leading journals and at several international conferences. He is the representative of INTERVAL (the Spanish Pole of the INTEROP-VLab) and a member of several research associations, such as EurOMA, POMS, IFIP WG 5.8 Enterprise Interoperability, and ADINGOR, among others. His key research topics include enterprise modeling, knowledge management, production planning and control, and supply chain management.

Martin Rudberg

Rudberg is the LE Lundberg professor of Construction Management and Logistics at the Department of Science and Technology at Linköping University in Sweden. He received an MS in Industrial Engineering and Management, and a PhD in Production Economics, both from Linköping Institute of Technology. Previously, Rudberg had been managing the Swedish Production Strategy Centre as well as large research projects on opera-  tions management in the process industries and on supply chain management employing advanced planning and scheduling systems. This included working with large Swedish-based companies, such as Alfa Laval, AstraZeneca, Ericsson, IKEA, and Toyota MHE. His main research interests include operations strategy, advanced planning systems, and supply chain management in the construction sector. He publishes his research regularly in leading journals.

Javier Santos

Santos is professor of Operations Management and head of the Industrial Management Department at the Engineering School, TECNUN, University of Navarra, Spain. He received his PhD in Industrial Engineering from the University of Navarra. He has been working as an industrial consultant and has directed more than 200 graduate master's theses related to Lean manufacturing and production plan-  ning and scheduling, which are his main research interest. Santos published a book entitled *Improving Production with Lean Thinking*, which has been translated into four languages.

chapter 1

Current Trends in International Operations

Migel Mari Egaña, Bart Kamp, and Ander Errasti

> *It's the possibility of having a dream come true that makes life interesting.*
>
> **Paulo Coelho**

Contents

1

Introduction

This chapter includes:

- Book content and purpose
- Concepts related to international operations
- Drivers of the globalization of operations
- Main reasons for starting up an internationalization process
- SMEs internationalization
- Future perspectives and trends

The phenomenon of globalization has skyrocketed in the last decades due to the opening of borders in Eastern Europe and the sudden emergence of other countries in the global trade economy. To be sure, growth in the European Union is one of the driving forces, but the main drivers have been the opening of borders in Asia—China being the prime example—to the global market economy and the relocating of production sites in order to enter into new trading blocks, such as NAFTA (North American Free Trade Agreement).

Internationalization of manufacturing and logistics networks is a phenomenon that has gained momentum over the past decade as a consequence of the evolution of the competitive environment. Many manufacturing companies, in fact, have increased their international presence to remain competitive.

The internationalization of operations can take different forms and includes the development of new configurations, such as international distribution systems, networks of global suppliers, and multisite and/or fragmented manufacturing networks.

Whether a company should begin the internationalization process is one of the most difficult decisions to be made because it involves a lot of risks, not only for multinational companies, but also and especially for SMEs (small and medium enterprises), which have limited resources, limited market knowledge, limited use of networks, and limited international experience of an entrepreneur.

Nowadays, the fragmentation of production processes and the multilocation of activities have gained great relevance. Companies are internationalizing more and more thanks to value chain configurations that extend beyond national borders. As a result, what is exchanged internationally are production tasks related to production operations, rather than assets. These changes have an impact on economies that experience the process of multilocation.

The internationalization process starting up is one of the most difficult decisions to make, because it implies a lot of risks not only for multinational companies, but also and especially for SMEs and Business Units from industrial divisions, whose resources are limited.

Many SMEs and Business Units in industrial divisions that enjoy local success have difficulty in foreign markets. Many of them are in the initial phase of the internationalization process in emerging market economies as well as in countries that are in transition, and they need to be prepared for the challenges of new market economies (Szabó, 2002). Therefore, dealing with the management of a production logistic network in different countries is an enormous challenge that requires a greater configuration and coordination efforts to get optimal levels of quality, flexibility, and cost (De Meyer et al., 1989).

This book based on the **GlobOpe (Glob**al **Ope**rations**)** framework **pretends to fill the literature gap in the implementation process of Global Operations.** Two perspectives will be adopted for this purpose. On the one hand is the production plant-level approach and on the other hand, the production network approach. The content of the book will look into:

- **Design and configuration** of Global Production Networks: frameworks, tools and techniques for aiding the location, design, and configuration of design, procurement, manufacturing, and distribution activities. Both are for production facility level and production network level including the suppliers' network.
- **Support tools to speed up** the new facilities, supplier network development, and multisite production network configuration **ramp-up processes**; risk analysis for setting up a new production plant overseas, methods to select alternatives, manufacturing strategic role definition and upgrading, adaptation of world class manufacturing methods and tools, best practices, etc.

This book points out the relevance in proposing the GlobOpe framework to accomplish three topics that are critical when configuring and managing a global production and logistic network (Figure 1.1):

- New facility implementation overseas (plant level)
- Global suppliers network configuration (both plant and network level)
- Multisite production configuration (network level)

It is designed to be an advisory tool for managers and practitioners who are in the internationalization process of their operations.

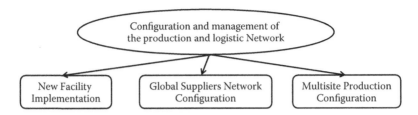

Figure 1.1 Reduced scheme of a GlobOpe framework.

Concepts Related to International Operations

Until recently, the domain or scope of Operations Management was domestic markets or regional areas. Most production processes were performed within the borders of a country. Although some producers imported their raw materials, all production and related finishing processes were executed within the same country. Moreover, the majority of the products were consumed locally, which meant managers were rarely involved in processes that went outside of their national borders.

In recent decades, this situation has changed dramatically. Many companies are suffering due to a trend toward greater fragmentation of their production and logistic processes. Thus, different value-added activities, such as engineering, purchasing, manufacturing, and assembly, are now undertaken at different sites, even in different countries.

The reduction in transport costs and the improvement of communications in general has facilitated the internationalization of manufacturing processes. Some recently industrialized countries provide resources (labor, in particular) at a much lower cost than traditional locations in developed countries. One of the causes is the possibility of easily moving (relocating) production facilities and expert labor, in addition to the elimination of trade and financial barriers within markets.

The following section includes several terms directly related to the internationalization of operations.

Globalization

Globalization has been defined in various ways. Two widely used definitions are:

1. "The process by which businesses or other organizations develop international influence or start **operating on an international scale**" (Oxford Dictionary).

2. "The increasing integration of economies around the world, particularly through the movement of goods, services, and capital across borders." The term sometimes also refers to the movement of people (labor) and knowledge (technology) across international borders. There are also broader cultural, political, and environmental dimensions of globalization (International Monetary Fund).

Therefore, we can say that globalization is closely related to the ability to develop and manufacture products in different regions other than in the company's country of origin.

The goal of globalization is to use the advantages it offers in terms of size and knowledge to produce additional sales in the new markets. Frequently, due to the costs of transferring products to distant places, the company is obligated to manufacture locally in the new market instead of in its own country.

Internationalization

The definition of internationalization according to Real Academia Española is

> Economical and entrepreneurial activities carried out by companies beyond their traditional geographical markets.

Therefore, the degree of a company's internationalization is determined by the number and proportion of activities that it develops in external markets. The goal of internationalization is not only to be bigger, but also better and more competitive.

For this reason, before embarking on the process of internationalization, a period of analysis and thinking is needed.

It should be remembered that internationalization is a means rather than an end in itself. Therefore, the authors of this book suggest that companies engage in deep and strategic thinking about future production and supply network architecture in order to gain a competitive advantage through the internationalization of their manufacturing.

Sales versus Domestic/International Production

Luzarraga (2008) explains that according to the volume of international sales and the international production, companies can be classified as local, offshore, export, or global companies (Figure 1.2).

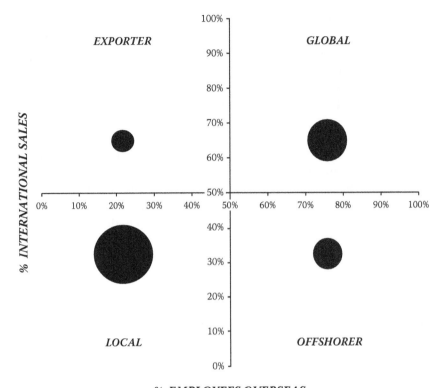

Figure 1.2 International strategy scenarios: sales versus production. (From Luzarraga Monasterio, J. M. 2008. *Mondragon Multi-Location Strategy—Innovating a Human Centred Globalisation.* Mondragon University, Onati, Spain.)

- **Local:** Sales are mainly focused on local or domestic markets where international sales are less than 50% of total sales, and the labor force outside the local or domestic border is less than 50% of the total labor. For most companies, this is the starting point when facing the internationalization process.
- **Offshore:** This strategy is based on a local market, with an international sales rate that is less than 50% of total sales, but where more than 50% of the labor force is located outside the local market. This strategy is the result of the special features of the industrial sector, the lack of competitiveness in sales because of the new competitors in low-cost countries, or the result of the offshoring process, which reduces costs and increases profit margins in locals markets.
- **Export:** The sales activity is global, with a percentage of international sales higher than 50%, but the labor force located abroad is lower than 50% of the total labor. Companies following this strategy have high

performance in international markets. More than a stage, this could be a natural second step in a company's internationalization process.

- **Global:** These companies have rates higher than 50% in sales as well as an international labor force over 50%. Such companies have reached a strategic position in sales in global markets and productivity. Hence, the market is seen as a single market and the companies can operate more effectively and better manage production footprint decisions (such as the supply network and manufacturing locations).

Employment Generated at the Headquarters and Abroad

In the process of internationalization, some authors (e.g., Luzarraga, 2008) argue that it is necessary to consider rates of employment growth at the headquarters and abroad (Figure 1.3).

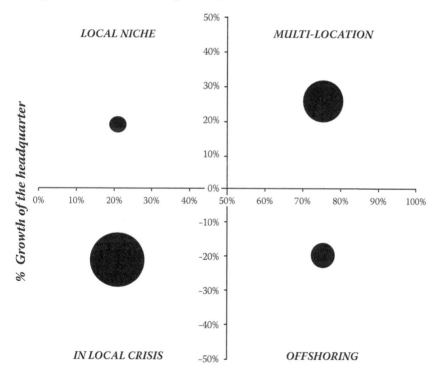

Figure 1.3 International strategy: Global production versus employment growth. (From Luzarraga Monasterio, J. M. 2008. *Mondragon Multi-Location Strategy—Innovating a Human Centred Globalisation*. Mondragon University, Onati, Spain.)

Following this criterion mentioned by Luzarraga, a company's employment strategy can be classified as:

- **Multilocation:** The globalization process not only generates new employment abroad, but there is also increased employment at the headquarters, especially skilled employment. This strategy is consistent with increased international sales, such as local sales.
- **Offshoring:** The globalization process generates new employment abroad and reduces employment at the headquarters. This strategy is the outcome of an offshoring process; it is consistent with a lack of international sales and a policy of outsourcing and externalizing operations because of a cost reduction or a dependence on foreign technology.
- **Local niche:** This kind of company doesn't generate employment abroad, but it is based in a local production niche, which means it generates new employment at the headquarters. The competitive advantage of these companies may not be due to the competitive cost of production, but rather to the incomes obtained with the product within a service more global than the company has got and that can offer to their customs.
- **In local crisis:** This kind of company is unable to generate employment abroad and can barely maintain employment at the headquarters. This is a troublesome state and the result of not having an international strategy. If there is not a change to the international strategy, the company will be in socio-economic difficulties.

Drivers of the Globalization of Operations

Even though there have been global operations since the Roman Empire, modern globalization has new drivers. Nevertheless, deindustrialization in developed economies has been of great concern since the 1980s.

In the current debates on globalization, controversy rages about the historical dating of the phenomenon. Nobel laureate Amartya Sen (2002) argues that globalization is at least a few thousand years old and that the West played a very minor role in its early phases. Sen rejects the commonly believed association between globalization and Westernization. At the other extreme, some scholars regard globalization as being a post-World War I phenomenon. Other researchers focus on events post-World War II. Economic globalization in the post-World War II era has been spurred by the successive rounds of trade liberalization under the auspices of the General Agreement on Tariffs and Trade (GATT), the forerunner to the World Trade Organization (WTO).

Some globalization skeptics argue that the Industrial Revolution was the breeding ground for globalization, while others point to the period of

European colonialism that dates from 1492, when Christopher Columbus first arrived in the Americas.

One implication is that the emergence of globalization coincides with the emergence of capitalism. For example, the establishment and expansion of the first global trade networks of the Dutch and English colonial trading companies would not have been possible without a system of capital reinvestment, private ownership, and commercial insurance.

Two critical landmarks, therefore, can be identified in the predevelopment phase of globalization:

- Explorers arriving in America, which symbolizes the emergence of colonialism.
- The emergence of the first multinational company.

Moreover, technological innovations, particularly those in transport and in communications technology, form a second primary foundation for globalization. According to Langhorne (2001), globalization originates in the second stage of the Industrial Revolution, with James Watt's pivotal invention of the steam engine in 1765. Langhorne distinguishes three phases of technological innovation that marked the process of globalization.

- The first phase is characterized by the application of the steam engine to land and sea transport and the invention of the electric telegraph. Steamboats and steam locomotives significantly reduced transportation time and increased transport volumes. This development increased the scope of industrial activities, thereby increasing the quantity of goods, the distances to which goods could be shipped, and people transported. It also made the distribution of information faster and less costly. The invention and improvement of the telegraph by Gauss, Weber, and Morse between 1830 and 1850 separated the speed of communication from traditional forms of transportation for the first time. The latter represents a historical turning point in the development of globalization, since distances in space and time decreased significantly.
- The second phase began during World War II when German engineers working on the V-2 project invented rocket propulsion. After the war, the intense technological competition between the Soviet Union and the United States accelerated the development of rocket and satellite technology. Thus, a truly global and reliable communication system was established for the first time in human history.
- The last phase is the invention of computers. Although the invention of the computer dates from as early as 1942, the capacities of the first computer barely exceeded the capability of today's hand-held calculators. However, the invention of the microchip in 1971 by Intel

increased the speed, processing volume, and efficiency of computers. Similar to the introduction of the electric telegraph, the invention of the microchip can be considered a major turning point in the development of globalization.

Another important technological development has been the innovation in transport technology, such as container transport and passenger aircraft. Since the end of World War II, the international mobility of people and the international tradability of varieties and quantities of goods have increased dramatically. Although the rapid growth of international passenger flights and transport increased over a longer time span, a concentration of growth can be discerned in the 1970s.

New Technologies

The advances in technology today are due to the Internet and information and communication-related technologies (ICT). Many analysts argue that ICT and its infrastructure (cable networks, satellites, etc.) have created a new techno-economic paradigm to the point that, as Kondratiev (2002) indicates, they have become the fifth growth cycle of an economy. Examples of ICT applications include:

- The EDI (Electronic Data Interchange) allows other enterprise systems (ERP, CRM, etc.) to connect so that information shared between companies can be processed without manual intervention, speeding the global exchange of information and documents for industrial information exchange and the exchange of trade, financial, medical, administrative, manufacturing, or other data, such as invoices, purchase orders, customs declarations, etc.
- B2B (Business to Business), B2C (Business to Consumer), and B2E (Business to Employee) are concepts related to communication practices and maintaining relationships with companies, customers, and the members of an organization. Apart from improving conventional operations by automating operations, eliminating errors, and increasing speed, electronic commerce is opening new ways of doing business.

The developments of overseas logistic operations, especially sea freight traffic, have contributed to the globalization. Some of the maritime traffic characteristics include:

- **Versatility and capacity:** Ships are highly specialized depending on the nature of the goods to be transported (bulk, container, etc.) and come in a wide range of sizes, from 100 DWT (deadweight tons) all the way up to 300,000 DWT.

- **International spread:** Shipping has the lowest average cost for transporting large volumes of goods between geographically distant points.
- **Poor traceability** in tracking merchandise.
- Relatively **low speed**.
- The need for **ground infrastructure and customs services** to improve the operating duty cycle.
- **Competence:** Despite some protectionist tendencies, most of the international traffic is carried in free competition according to the laws of the freight market. Freight is the price of the shipping service.

For such maritime transport, it has undergone a process of concentration and there has been growth in main harbors and transport hubs. These harbors have large port facilities for storing and unloading goods and connecting to land transport, which has allowed vessel size and cargo capacity to grow.

The standardization of load units (containers) has facilitated the proliferation of items and goods that can be transported overseas (Figure 1.4).

Many ports could be simply dismissed because they are not large enough or because they lack the necessary infrastructure to accommodate container ships.

Economies of scale generate an increase leading to an increase in vessel size. This can be observed in Figure 1.5.

Figure 1.4 View of a container terminal in a harbor.

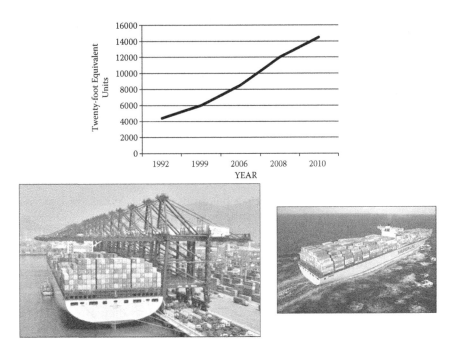

Figure 1.5 The *Emma Maersk* is currently the biggest container ship in the world.

Trade Regulations

Recently there have been changes in trade regulations that affect internationalization. The most significant ones include:

- The creation of the European Economic Union's Treaty of Maastricht (1993), which allows free commerce, the introduction of the Euro as the single currency (1999), and the growth of the European Community through the addition of more countries, particularly in Eastern Europe.
- The fall of communism in Europe in the 1990s has been one of the biggest changes in the global order. This brought the unification of Germany and the extension of the capitalist system to the countries in Eastern Europe, some of which have been integrated (or they will be in a short time) in Western political institutions (e.g., the European Community, NATO). The growth expectation for Eastern Europe combined with the low costs of labor is causing an avalanche of new companies to locate in countries such as Poland and the Czech Republic.

- Moreover, free trade areas are being created in other parts of the world: NAFTA (the North American Free Trade Agreement, an agreement between the United States, Canada, and Mexico), AFTA (the Asian Free Trade Agreement, an agreement between the countries of Southeast Asia), and similar movements in South America. Due to the advantages (labor cost, raw material cost, etc.) that these free trade areas offer to companies, it may be beneficial to establish production plants in them.

Another country to consider due to its importance in international trade is China. It has been evolving from a centralized planned economy to a market economy. The enormous potential of this country and the different alternatives as a new market (purchasing that it offers) have led to one of the biggest Foreign Direct Investment (FDI) movements.

In China, as well as in other low-cost countries, the attitudes of state or regional governments toward private property, urbanization, pollution, and employment stability may vary. The attitudes of governmental agencies may not last long when the governmental agencies have to make a decision on a location. Moreover, a company's management may find that these attitudes are influenced by the party or person in power (Political Risk).

Culture

During the last years of the last century and the beginning of this century, the movement of people has increased around the world. Open borders and the increased ease of travel are the reasons for the increasing number of immigrants (legal as well as illegal). Therefore, people are moving greater distances and going to more exotic places as tourists. Many young people choose to study abroad, as well as spending periods of time abroad before finding a steady job. Satellite television and the Internet make it possible to examine and learn different lifestyles.

People want to have products and services that are available in the most advanced economies, encouraging the trend toward the use of global product. These products have the same quality independently of the place where they are produced or consumed. That is to say, market demand has homogenized and the number of different markets that accept the same product or at least products with low differentiation is increasing. Ohmae (1985) calls this phenomenon *Californization*, and it allows companies, such as Coca-Cola, McDonalds, and Toyota, to sell the same products in virtually any country.

The attitude of workers may vary from one country to another, from one region to another, and from one city to a metropolis. The opinions of workers regarding the performance, unions, and absenteeism are all relevant factors.

Economical Factors

While developed zones (the United States, Europe, and Japan and their areas of influence) have been enjoying more or less stable economic development since the end of World War II, in the last decades there has been a series of events that have affected in large measure the economic development of these countries. Some examples are the following:

- **The emergence of newly industrializing economies,** the most important of which are the Asian Tigers (South Korea, Taiwan, Singapore, Hong Kong, Malaysia, Thailand, the Philippines, and Indonesia), Latin America (especially Brazil and Mexico), and India. It should be noted that China is already considered as the "world's factory" and that it has become the fourth economic power in the world (BBC, 2005).
- **The openness of communist countries** to the market (e.g., Eastern Europe and Asia) has made them attractive firstly because of the development opportunities they offer as well as the low cost of resources compared to developed countries.
- **The role played by multinationals in the empowerment of globalization.** Some of these worldwide organizations can benefit the current financial and banking systems, which operate electronically 24 hours a day and allow capital to be shifted instantly and without restrictions from one point of the planet to another. Multinationals can make decisions regarding the location of their operations and these decisions can affect their potential suppliers and the economy of the country in which they operate. They, therefore, are powerful economic forces that influence policies and strategies of many companies (including governments).

Economists usually use per capita income as an indicator of a country's prosperity, where growth is determined by employment, the number of hours worked, and productivity.

The importance of productivity, which measures the production value per employee or, more frequently, per hour worked, is crucial in achieving greater competitiveness. A country's macroeconomic level depends on the number of people working and on this work being done efficiently with strong endowments of physical, technological, and human capital.

From the Operations Management point of view, all of these factors are translated at the microeconomic level of companies to factors affecting productivity.

Moreover, manufacturers looking for production economies of scale create factories whose production is higher than the local market demand. This paradigm forces a search for new markets.

The increased demand in new markets according to the regions raise their prosperity and living standards, besides foreign companies identify opportunities to sell into new markets.

In making decisions regarding production location, the following factors are usually taken into account as those affecting productivity:

- Hourly Rate (Euro/hour): Figure 1.6 shows the large difference in average monthly rates (minimum wage per month) for different countries and the data from the European Union, which were published in January 2009 by Statistical Office of the European Communities

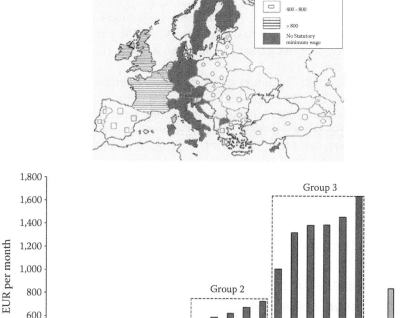

Figure 1.6 Minimum wages in euros (EUR).

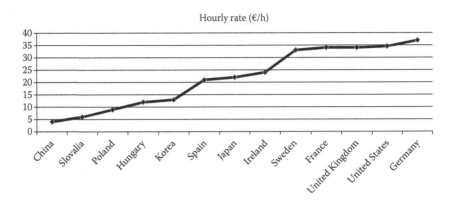

Hourly rate (€/h)

Figure 1.7 Hourly rate study.

(Eurostat). Group 1 is for minimum wages below EUR 400, Group 2 for EUR 400 to 800, and Group 3 for more than EUR 800. Finally, Figure 1.7 shows a graph where the hourly rate (€/h) for different countries is represented. The difference is obvious. For example, the difference between Germany and the Czech Republic is significant. The hourly rate in Germany is at €37.67 while in the Czech Republic it is just €10.

- Hours of work: The hours spent at work are different in each country. In 2011, the Organisation for Economic Co-Operation and Development (OECD) just released its study on average annual hours worked per worker in 2008, showing in which countries workers work longest (Figure 1.8 and Figure 1.9). At the economic level, there are certain factors that affect productivity and they are monitored, but there are others which can affect production efficiency that are not discussed by academics in depth (Barnes, 2002). These are the units produced per hour, or work efficiency, a factor that joins aspects of design and production and logistics processes. In addition, human capital can influence the efficiency as much as or more than the others.

Main Reasons for Starting Up the Internationalization Process

There are four main reasons why a company may want to start up the internationalization process, all of which have been explained by authors such as Ferdows (1997) and Farrell (2006):

- **To have production abroad or offshore:** The entire production process is placed in a new location. The competitive advantage of the new facility is primarily cost. From this location, the suppliers

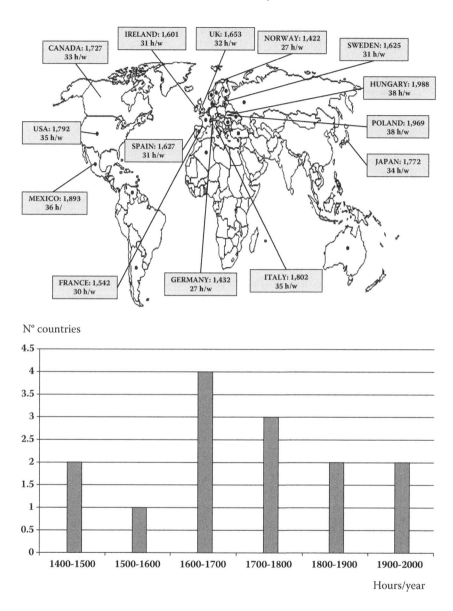

Figure 1.8 Map showing the hours of work per year and per week.

export products to old markets and new possible areas. Many low-cost countries in manufacturing have already achieved sufficient technological development to bring in the production supply chain of almost all industrial sectors. However, it is necessary not only to scrutinize the production costs, but also to see the difference

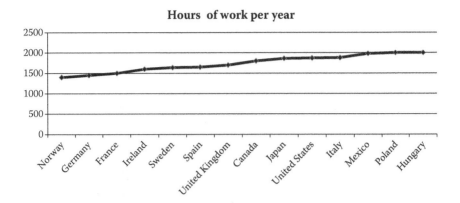

Figure 1.9 Graph showing the hours of work per year.

between production costs at the production place and prices in export markets—a calculation that is more appropriate than just looking at labor costs. There are other cost factors that may have as much or more importance such as land costs, taxes, etc.

- **To enter new markets:** Companies enter new countries to expand demand and their consumer base. The new facility is needed to enter this market because there is a trade barrier or a transport/ logistic lead time that keeps this market from being supplied from the matrix. The production model in the foreign country tends to replicate the one used at the matrix. Despite growing in sales in other markets, the factor of entering new markets can mean facing high entry barriers: tariffs, distance, cost-effectiveness (Jarillo and Echezarraga, 1991). The best competitive position is one that allows a company to operate without rivals within the major markets (which are continuously changing and evolving) and to have accessibility to other markets with few restrictions (Lafay and Herzog, 1989). Thus, the deeper the integration and location of a company's production and logistics activities in a specific economic area are, the more opportunities in some cases to increase productivity or to provide a response of value-added because of its service and quick response (agility in the supply). Therefore, the trade dynamics and foreign investment between countries and major regions decisively affect the performance of companies or company networks.
- **To disaggregate the value chain and re-engineer the value chain:** The supply chain is fragmented and located, centralized or decentralized, in different locations or regions. The manufacturing and supply network task and processes are redesigned to take advantage of each facility individually (Ferdows, 1997) and to obtain synergies from the coordination of this network (Mediavilla and Errasti, 2010).

To create new products and markets: By capturing the full value of global activities and using the capabilities of the manufacturing network, companies can offer new products at significantly lower prices and penetrate new market segments. This concept should include the scientific basis for the production and management process, the importance of R&D (research & development), human resources for technological innovation, the appropriate use of new technologies, and the degree of diffusion in the entire network of economic interaction. In other words, technological capability is not only the sum of several elements, but also an attribute to be designed into the system of production and logistics. It refers to the perfect union of science, technology, management, and production in a complementary system, whose levels are provided by the educational system, with human resources necessary for training and producing sufficient quantity. The competitiveness of OECD countries is determined by the technological level of each sector. Similarly, the ability of a country to fight in the international economy is directly related to its technological potential.

Some authors identify a number of other reasons to begin the process of internationalization and these include:

- **To diversify risk** between different countries (Thompson and Strickland, 2004) and compensate for temporary losses in some regions with the gains in others (Jarillo and Echezarraga, 1991).
- **To achieve economies of scale** in production activities, R&D, distribution and procurement, among other things, which allows for a reduction in costs (Deresky, 2000).
- **To capitalize on major investments** (Jarillo and Echezarraga, 1991) and compensate for the reduced cycle life of products with exploitation on a larger scale by taking advantage of the relative homogeneity of markets (Yip and Bink, 2007).
- **To acquire prestige**, continue to grow globally, and gain competitiveness (Jarillo and Echezarraga, 1991).

There are additional reasons, which can be seen in Figure 1.10.

SMEs Internationalization

The research regarding SMEs internationalization began in the early 1970s in the Nordic countries and produced gradualist models (Cavusgil, 1980; Johanson and Vahlne, 1977). The main output is the Uppsala model (U-model) (Johanson and Vahlne, 1977, 1990, 2003). The U-model assumes that the internationalization process begins according to the level of knowledge of the market. Usually, it begins with sporadic overseas sales and continues with increasingly larger and

Why manufacture Abroad?

Most tangible

Reduce direct and indirect costs
Reduce capital costs
Reduce taxes
Reduce logistics costs
Overcome tariff barriers
Provide better customer service
Spread foreign exchange risks
Build alternative supply sources
Preempt potential competitors
Learn from local suppliers
Learn from foreign customers
Learn from competitors
Learn from foreign research centers
Attract talent globally

Most intangible

Figure 1.10 Reasons for the internationalization process.

gradual commitments in the foreign markets through sales; the more they learn, the higher the commitment, and the higher the commitment, the more they learn. Another characteristic of the internationalization process is that it is usually initiated in the markets perceived close to the domestic one. Due to the liability of foreignness (Hymer, 1976; Zaheer, 1995), a firm starts internationalization in countries that are perceived culturally, economically, geographically (etc.) close before gradually entering other markets (Johanson and Vahlne, 2009). The start-up of an overseas production unit usually necessitates deep understanding of the local market.

Then, in the 1990s researchers found that some SMEs were able to internationalize more rapidly than the gradualist models predicted (Oviatt and McDougall, 1994, 2005). Therefore, the theories of International New Ventures (INV), "born global," or "born-again global firms came up. The first two types of firms are internationally oriented and have been since their inception or soon after, and they manage to reach a certain degree of internationalization within a relatively small number of years (e.g., three, five, or six) (Bell et al., 2003). On the other hand, born-again global firms operated for a number of years only on a national base and, due to a critical event, changed the strategy and internationalized rapidly.

The two key sources in such a rapid internationalization process are knowledge and international networking (Oviatt and McDougall, 2005):

- Firms with **higher market knowledge** (e.g., due to the entrepreneur's international experience) have a higher propensity (or learning capability) to gather further foreign knowledge (Oviatt and McDougall, 2005), and the knowledge intensity develops the learning skills and makes it easier for firms to adapt in a new environment (Autio et al., 2000).
- An **international network** helps the entrepreneurs in spotting the opportunities, establishing international relationships, and accessing information.

Table 1.1 describes and compares the differences between both approaches (Gradualist and INV), focusing on the framework for the internationalization pathways developed by Rialp et al. (2005).

Recently, Johanson and Vahlne (2003, 2009) also noticed the importance of international networking while revisiting their model. They argue that the major obstacle in internationalizing no longer consists of the liability of foreignness but instead in the liability of outsidership, i.e., being or not being part of a network that makes a difference. Nevertheless, current researches (Kalinic and Forza, 2011) have found that the SMEs can overcome the liability of outsidership by developing the network during the internationalization process itself through integration on board of (sometimes) unexpected stakeholders (Sarasvathy, 2008). In other words, an international network is not necessarily a precondition.

Furthermore, Kalinic and Forza state that SMEs are able to speed up their internationalization process. They suggest that specific strategic focus (as opposed to knowledge-intensity, international network, and international experience) is the determinant success aspect of the change in the internationalization process that allows the traditional SMEs to rapidly internationalize the operations in unknown markets. More specifically, the persistent endeavors to form local relationships, the proactive entrepreneurial orientation in a host environment, and a flexible strategic focus with heterogeneous expectations positively affect respectively the extent of international commitment, the scope of international commitment, and the development of commitment in the host country.

Perspectives and Trends with Regard to the Internationalization of Business

In the following sections, we highlight a number of prominent trends with regard to actors and factors in the internationalization of business in the recent past and the years to come.

Table 1.1 Differences between born-globa/INV theory and gradualist model's firms

Key Dimensions	Attribute	Born-Global/INV Theory	Gradualist Model's Firms
Founder's (and/or founding team's) characteristics	Managerial vision	Global from inception	International markets to be developed gradually after a significant domestic market base
	Prior international experience	High degree of previous international experience on behalf of founding entrepreneurs and/or managers	Irrelevant or low degree of previous experience in international issues
	Managerial commitment	High and dedicated commitment with early internationalization efforts and challenges	General commitment with objectives and tasks, but not directly related to internationalization
	Networking	Stronger use of both personal and business networks at the local and international level Crucial to firm early, rapid, and successful global market reach	Loose network of personal and business partners; only foreign distributors seem to be relevant to the firm's gradual path and pace of internationalization
Organizational capabilities	Market knowledge and market commitment	High from the very beginning due to superior internationalization knowledge at inception	Slowly growing with previously accumulated domestic and foreign market knowledge
	Intangible assets	Unique intangible assets (based usually on knowledge management processes) are critical for early internationalization purposes	Availability and role of intangible assets are less important for successful gradual internationalization

(continued)

Table 1.1 Differences between born-globa/INV theory and gradualist model's firms (continued)

Key Dimensions	Attribute	Born–Global/INV Theory	Gradualist Model's Firms
Organizational capabilities (cont.)	Value creation sources	High value creation through product differentiation, leading-edge technology products, technological innovativeness, and quality leadership	Less innovative and leading-edge nature of its products resulting in a more limited value creation capability
Strategic focus	Extent and scope of international strategy	A niche-focused, highly proactive international strategy developed in geographically spread lead markets around the world from inception	A more reactive and less niche-focused international strategy International markets will, at best, be developed serially and in order of psychic distance
	Selection, orientation, and relationships with foreign customers	Narrowly defined customer groups with strong customer orientation and close or direct customer/client relationships	In the hands of intermediaries at the earliest stages of internationalization
	Strategic flexibility	Extreme flexibility to adapt to rapidly changing external conditions and circumstances	Limited flexibility to adapt to rapidly changing external conditions and circumstances

Source: Rialp et al. 2005. The born-global phenomenon: A comparative case study research. *Journal of International Entrepreneurship* 3 (2): 133–171. With permission.

Change in the Geographical Point of Gravity for Business Hotspots: From Triad Power to BRIC Power

At the end of the 1980s (Ohmae, 1987), 70% of the world GDP (gross domestic product) and 75% of world trade was concentrated among the triad power nations (Europe, the United States, and Japan). During the past three decades, this picture has severely changed. BRIC (Brazil, Russia, India, and China) countries represent an increasing and quickly growing share in the global GDP, and, in the slipstream of these countries, there are also a number of other countries that are increasing their footprint on global business rapidly and substantially (e.g., Hong Kong, Singapore, South Korea, and Taiwan). BRICs have moved from 11% of GDP (about 30% for emerging markets overall) in 1990 to 16% in 2000 (37% for emerging countries) and around 25% currently (50% for emerging markets). Moreover, this tendency is expected to continue with BRICs reaching about 40% of global GDP by 2050 (73% for emerging markets). Moreover, in 2010, China was already the second largest economy in the world, Brazil 7th, India 10th, and Russia 11th. By 2050, the BRIC economies (Wilson et al., 2011) are projected to make up four of the five largest economies in the world: China in first place, India in third, Brazil in fourth place, and Russia in fifth place, with the United States in second place to complete the top five. Furthermore, BRIC countries are becoming increasingly important players in global trade flows. While in 2000, these countries accounted for 6% of global trade flows, this amount increased to 15% by 2010 (Goldman Sachs, 2011).[*] In this regard, not only are these countries becoming increasingly important trade partners for the traditional economic powerhouses (the Triad), in addition (e.g., through an intensification of trade relations between Asia with Latin America and Africa), it is becoming more and more apparent that economic ties between emerging continents and countries are growing and that this reshapes the course and intensity of global trade relations overall (King, 2011). In the slipstream of this process, the pivotal points of global trade also shift and ports like Shanghai and Singapore emerge more and more as leading global hubs. Finally, also, geographical patterns of FDI flows are changing. In 2010, and for the first time in history, developing economies absorbed about half of the global FDI flows (UNCTAD, 2011). That year, they also generated record levels of FDI outflows, representing 29% of the global FDI outflows, much of that share taking place between countries in the southern hemisphere.[†]

[*] Goldman Sachs (2011). *The BRICs 10 years on: Halfway through the great transformation.* Global Economics Paper No. 208.

[†] UNCTAD (2011), *World Investment Report 2011.* Non-equity modes of international production and development. United Nations, New York and Geneva. Online at: http://www.unctad-docs.org/files/UNCTAD-WIR2011-Full-en.pdf

Change in the Business Functions That Are Being Off-Shored: From Production and Sales to Innovation and R&D

It is becoming increasingly common that companies do not go abroad only for tapping into lower costs of production factors or for the fact that foreign markets grow faster. In addition, more and more companies turn to foreign markets as stepping stones for innovation. This happens especially with countries that form fertile grounds for new product ideas or new ways of conceiving and delivering products to market. Notably when they shelter vast amounts of ingenious and curious consumers, albeit with lower purchasing power standards, which trigger the need for disruptive innovations and to overhaul occidental product design and market penetration schemes, these countries are interesting for new product development.

Whereas in the West, new products are typically first introduced for vanguard and high-end consumers (as those are the most experimental or least-demanding ones), only to mainstream such products in a second stage. In many emerging countries, the most experimental consumers are often situated in the lower echelons of the market pyramid, and it is these consumer groups that are most inductive to innovations.

In a double sense then, those customer segments can function as a trampoline for what is often called *reverse innovation* (Govindarajan and Ramamurti, 2011). Reverse in the sense of developing innovations and new products first for entry level consumers and later on to adapt them for the most sophisticated users. And reverse in the sense of first for emerging markets and afterwards for the so-called advanced economies. The latter embodies also a rupture with the classical international product life cycle thinking (Vernon, 1966), which argues that products first serve Western markets and once demand gets saturated there, companies will look for less advanced countries to offer their products there.

Among others, companies like GE developing a handheld electrocardiogram apparatus in India, and Vodafone pioneering with mobile phone money transfer applications in Kenya, have had success stories with this formula. In furtherance to practices in Africa, India is now also leapfrogging from "no landline telephones" to wireless coverage of its territory. That enables the country to implement wireless banking practices for rural and peripheral populations that have no access to (nearby) banking offices.

These cases illustrate the fact that companies can get to a point where they turn foreign markets not only into places for production, but also into loci for innovation activities. On the one hand, this may be due to the fact that they may—in terms of available engineers, cost-wise, and idea-wise—be very fertile places. On the other hand, and especially if products are not commodities or bulk issues, this may stem from the need for customization and local responsiveness, which also induces the set-up of continent-specific innovation/R&D structures.

This also implies that entire value chains become embedded in emerging countries and the risk of foot looseness of such industrial complexes in Western countries increases (*cfr.* as Steve Jobs argued to Barack Obama in February 2011: "Those jobs aren't coming back"). By means of illustration of the former, seminal technologies and disruptive innovations often pave the way for new product lines or for radical makeovers of existing product lines. If such new inventions and innovations are pioneered in overseas markets, it often gives production facilities in those places an advantage over others in the roll-out and scaling up of serial uptake of new technologies. Alternatively, it enhances the probability of opening a pilot production plant in other countries to take charge of this. For there is typically a close link between the R&D part of a company, where technologies and proceeds are invented and tested on a small scale, and the (pilot) plant(s) that take up the findings. Moreover, lab researchers are often part of the project team for rolling out and scaling up the new technologies and techniques. With the tendency to delocalize innovation and R&D functions abroad, the tendency to delocate production becomes even more fortified (see, for example, experiences in the chemical industry, like from Evonik and Degussa, 2012).

Change in Governance Modes and Timing of Foreign Market Entry: From Internationalization as an MNE-Only Game to the Rise of Born Globals

Extant theories on internationalization processes and the rise of multinational enterprises (MNEs) argue that going abroad is typically an affair for large scale and vertically integrated companies that can take advantage of managing assets spread across borders through internalization skills (e.g., Dunning, 1980, 1988a, 1988b, 1992). Similarly, the dominant view on how companies internationalize is that they first go to (geographically and/or culturally/linguistically) nearby countries[*] only to take on more distant markets once they have acquired a certain level of experience in dealing with foreign countries (Davidson, 1980). In a similar vein, it is posited that companies first internationalize their business via low equity/commitment entry modes into foreign markets (exporting, licensing/franchising), only to shift to more resource-intensive entry modes (such as joint ventures, foreign acquisitions, and opening of wholly-owned subsidiaries and greenfield investment) in a second instance. These theories put emphasis on the incremental and gradual internationalization of firms through the acquisition, integration, and use of knowledge concerning foreign markets and on a successively increasing commitment to those markets over time.

[*] Some authors introduce the concept "psychological distance" to refer to the degree of differences in aspects such as language, culture, policy, educational level, or industrial development (e.g., Johanson and Vahlne, 1977; 1990).

However, there is a growing body of evidence on small technology-oriented firms (formed by active entrepreneurs and often emerging as a result of a significant breakthroughs in specific process or technology areas: McKinsey & Co., 1993; Madsen and Servais, 1997; Gassmann and Keupp, 2007) that operate internationally almost from inception (exporting more than 25% of sales within three years after start-up). Myriad examples from across the globe can be found in this regard. One of them is the Basque firm Salto Systems, devoted to the intelligent locking solutions. Established in 2001, the company found its market niche in a business dominated by strong multinationals. Still, five years after its creation, the company was already established in about 60 countries. At present, more than 90% of the company's revenues come from foreign markets, and the company has offices in the United Kingdom, the United States, Canada, Mexico, Portugal, Australia, the Netherlands, the United Arab Emirates (UAE), and Malaysia.

The existence of these so-called born globals thus illustrate that the patterns as portrayed in the works of Johanson and Vahlne (1977, 1990) are largely overtaken by reality or at least are not universally applicable anymore (Chetty and Campbell-Hunt, 2003).

According to Knight and Cavusgil (1996), several factors have given rise to the emergence and success of these born global firms. For example, niche markets, in which smaller firms can specialize and make their mark more easily, play an increasing role on the global economic scene and allow (and oblige) firms to gather clients from multiple marketplaces. Second, advances in process technology enhance the profitability of production on a small scale. This also raises the possibilities of making the customization of products to the tastes of buyers from different places cost-effective. Third, with the advances in communications technology, managers at small firms also can efficiently manage operations across borders. Furthermore, factors such as globalization on a whole and a more international outlook among entrepreneurs across the board favor the rise of the born global phenomenon (e.g., Andersson and Wictor, 2003).

Change in the Dominance of Global Industries: From Western Hegemony to Market Leadership on Behalf of Emerging Economy Giants

While globalization has opened new markets to rich-world companies, it also has given birth to a pack of fast-moving, sharp-toothed new multinationals that are emerging from the poorer world countries.[*]

---

[*] Globalisation's offspring. 2007. *The Economist* April 4.

The rise of new corporate powers from emerging countries is partly intertwined with increasing consumption and production activity in those countries, but, in a number of cases, can also be attributed to (technological) innovations. Two points in case are Mittal (Luxembourg) and Tata Steel (India), which became global leaders in the steel industry. Due to their involvement in the invention and perfecting of minimill technology, allowing quality steel to be made from scrap, they became masters of a so-called disruptive technology (Christensen, 1997). When it became clear that it could outperform iron ore-based steel production on cost and could attain an increasing range of qualities, minimill technology was mainstreamed and those in command of it turned the power balance in the industry. As a consequence, Mittal de facto took over Arbed/Arcelor, and Tata acquired British Steel and Hoogovens.

The accumulated purchasing of products made in emerging economies also led to the fact that a number of autochthonous companies developed extremely deep pockets. As a consequence, there have been cases where Western companies with cash problems ended up in the hands of companies from BRICs, like Jaguar and Land Rover being absorbed by Tata, and Volvo falling in the hands of Geely (China).

The former examples serve to illustrate the rise of companies from emerging companies, which is not anecdotic. In fact, it is a broad movement. Consider, for example, that in 1995 only 5% of companies composing *Fortune Global 500* were from emerging companies. Ten years later, this participation has doubled, and only five years more were needed to reach 20%. Consequently, in 2010, some 92 out of 500 companies came from Brazil, India, Mexico, Russia, South Korea, Taiwan, or China; the latter country providing most of the top-sized companies to the Fortune 500.[*] Especially in the oil sector, companies from countries like China (but also Brazil [Petrobras]) have been gaining positions. Take, for example, Sinopec (China), which was the first company that broke the Western dominance in *Fortune's* Top 10, rising to a ninth position in 2009 and to fifth in 2011. Meanwhile, also China National Petroleum and Chinese State Grid entered the top 10, testifying to the strength of China in a strategic sector, such as energy. Another industry where dominance has changed significantly is banking. Probe of this shift is that the two biggest banks in the world by market capitalization at the end of 2010[†] were two Chinese banks: Industrial Commercial Bank of China and China Construction Bank. Besides the Chinese and Indian examples, there is a growing number of megacompanies in South America starting to make their mark in strategic sectors of the globe. Brazilian Petrobras, for instance, was in

[*]   *Fortune Global 500*: http://money.cnn.com/magazines/fortune/global500/2011/index.html
[†]   Source: *Financial Times Global 500.*

seventh position in terms of market capitalization in 2010.* Moreover, Brazilian Vale (mining and energy) and JBS (retail) occupied the fourth and eighth place as regards revenue growth in *Fortune's* corresponding global overview for 2010.

As well as the trends mentioned above, there are others that will have a significant impact on the context for global manufacturing, such as (Christodoulou et al., 2007):

- **Environmental sustainability rises in corporate priorities:** The environment could theoretically create a reverse trend in offshoring. Direct fuel cost increases, taxes, and consumer power are all helping to make global sourcing and transportation between continents less and less attractive.
- **Technologies for distributed manufacturing are appearing:** More and more research is centered on flexible manufacturing solutions that require lower economies of scale. This, combined with increasing transportation costs, could present a tipping point where local production suddenly becomes much more attractive.
- **Servitization:** More and more companies are bundling services with products and engaging in life-long product support to maximize consumer intimacy and access higher margin activities. This leads to more customization and complexity in supply chains, which in turn demands new manufacturing capabilities together with responsive supply models (see Chapter 3).
- **World resources are increasingly scarce:** Declining reserves of primary raw materials and other key resources, such as water, are predicted to have an increasing impact on the conditions for global business. This leads to an increase in the refurbishment or "remanufacturing" of products and clearly affects the network of activities involved in producing new products.

Case: The integration of Fabricacion de Automoviles S.A. into Renault's global automotive production network

When Spain decided to develop a domestic automotive industry after World War II, it had to look for foreign firms that had the necessary technological expertise and financial resources to get this on track. As part of its autarchic stance at the time it did not allow foreign firms to set up factories of their own and/or assemble cars on the basis of imported parts only. As a consequence it imposed severe local content rules and intended

---

* *Businessweek* and Boston Consulting Group, 2009.

to team up foreign automotive companies with local entrepreneurs (e.g. Barreiros, Huarte) or industrial structures (e.g. SEAT). Renault of France was one of the foreign automotive companies that was lured to Spain where it engaged in a joint venture with "Fabricacion de Automoviles Sociedad Anonima" (FASA) in Valladolid.

At inception, the venture was allowed to temporarily import from France a significant share of the components needed to mount the cars. Albeit, at the condition that a swift build-up of internal capacity to produce these parts at FASA itself would follow. Consequently, Renault helped the FASA factory to internalize a sizeable amount of the parts production to be able to cease receiving excessive supplies from France. In the second half of the 1960s (when Spain adopted more liberal policies towards local content rules), FASA integrated gradually into the overarching Renault hierarchy. This process had its starting point in 1965, when FASA changed into FASA–Renault. Later on, from the end of the 1970s onwards, a further integration of Renault's foreign assets took place leading to increased international interactions between the French Renault apparatus and the foreign affiliates in terms of joint production planning and exchange of components manufactured at its respective plants.

Accordingly, from a geographic perspective, the supply relations of FASA can be characterized by a number of stages as well. At the outset it relied primarily on inputs from France and the Valladolid plant had a kind of stand-alone character. Afterwards, and in order to live up to its local content obligations; on the one hand it intended to internalize as many part production tasks as possible, and on the other it tried to complement this selectively with sourcing from Spanish suppliers –some of which were acquired if they were of strategic importance (like ISA of Seville). The outcome was a highly integrated production apparatus, complemented with selective sourcing of parts and components from third party suppliers. In spatial terms its production and sourcing apparatus stretched out on the axis Seville-Valladolid-Euskadi (the Basque Country) and Cantabria, where many of its parts suppliers were located. As it relied on extant companies and these were mostly located in places with a longer metal and mechanical tradition (the North of Spain), for a long time FASA (-Renault) would not produce many auxiliary spin-offs in its vicinity. In fact, as far as supply companies were set up in the region where FASA was located, these were mostly dedicated to less high-tech parts and pieces.

Later on, when FASA-Renault became consolidated in the overall Renault hierarchy, a considerable part of its independent suppliers from Spain were replaced by supply of parts and components from other Renault factories with superior economies of scale. In parallel, under impulse of Renault's purchasing direction, a process of thinning out the overall supplier base to Renault's factories was set in motion, which tended to favour the engaging of French suppliers. The latter were also induced to take-over traditional suppliers to FASA in order to enhance their logistical efficiencies vis-à-vis the respective assembly plants they would attend. This helped partially to consolidate the Spanish part of FASA's supplier base, while at the same time it meant a continuation of rather spread-out supply relations in Spain, without a concentrated supplier base inside Castilla y Leon (where Valladolid is located).

References

Andersson, S., and Wictor, I. (2003) Innovative internationalisation in new firms—Born Globals the Swedish case. *Journal of International Entrepreneurship* 1 (3): 249–276.

Autio, E., Sapienza, H. J., and Almeida, J. G. (2000) Effects of age at entry, knowledge intensity, and imitability on international growth. *Academy of Management Journal* 43 (5): 909–924.

Barnes, D. (2002) The complexities of the manufacturing strategy formation process in practice. *International Journal of Operations & Production Management* 22 (10): 1090–1111.

BBC. (2005) Zhongguo huo shijie gongchang touxian (China gains a factory of the world title). BBC Chinese website, Dec 16. http://news.bbc.co.uk/chinese/simp/hi/newsid_4530000/newsid_4535500/4535526.stm Retrieved on Dec 20, 2006.

Bell, J., McNaughton, R., Young, S., and Crick, D. (2003) Towards an integrative model of small firm internationalization. *Journal of International Entrepreneurship* 1 (4): 339–362.

Cavusgil, S. T. (1980) On the internationalization process of firms. *European Research* 8 (6): 273–281.

Chetty, S., and Campbell-Hunt, C. (2003) Paths to internationalisation among small- to medium-sized firms: A global versus regional approach. *European Journal of Marketing* 37 (5/6): 796–820.

Christensen, C. M. (1997) *The innovator's dilemma: When new technologies cause great firms to fail*. Harvard Business School Press, Boston.

Christodoulou, P., Fleets, D., Hanson, P., Phaal, R., Probert, D., and Shi, Y. (2007) *Making the right things in the right places. A structured approach to developing and exploiting "manufacturing footprint" strategy*. IFM, University of Cambridge, U.K.

Davidson, W. H. (1980) The location of foreign direct investment activity: Country characteristics and experience effects. *Journal of International Business Studies* 11 (2): 9–22.

De Meyer, A., Nakane, J., Miller, J., and Ferdows, K. (1989) Flexibility: The next competitive battle the manufacturing futures survey. *Strategic Management Journal* 10: 135–144.

Deresky, H. (2000) *International management: Managing across boarders and cultures*, 3rd ed. Prentice Hall, Upper Saddle River, NJ.

Dunning, J. H. (1981) International production and the multinational enterprise. London: Allen & Unwin.

Dunning, J. H. (1988a) The eclectic paradigm of international production: A restatement and possible extensions. *Journal of International Business Studies* 19 (1).

Dunning, J. H. (1988b) *Explaining international production*. Unwin Hyman, London.

Dunning, J. H. (1992) *Multinational enterprises and the global economy*. Addison-Wesley Publishing Company, Wokingham, U.K. and Reading, MA.

Evonik Degussa. (2012) Experiences from the chemical industry. http://www.pro-inno-europe.eu/inno-grips-ii/blog/domestic-rd-activities-are-base-future-chemical-plants-europe (accessed Feb. 15, 2012).

Farrel, D. (2006) *Offshoring. Understanding the emerging global labor market*. McKinsey Global Institute, Harvard Business School Press, Boston.

Ferdows, K. (1997) Making the most of foreign factories. *Harvard Business Review*, March-April: 73–88.

Financial Times (2001) FT 500: The world's largest companies. Online at: http://www.ft.com/intl/reports/ft-500-2011

Fortune (2011) Fortune Global 500: Online at: http://money.cnn.com/magazines/fortune/global500/2011/index.html

Gassmann, O., and Keupp, M. M. (2007) The competitive advantage of early and rapidly internationalizing in the biotechnology industry: A knowledge-based view. *Journal of World Business* 42: 350–366.

Govindarajan, V., and Ramamurti, R. (2011) Reverse innovation, emerging markets, and global strategy. *Global Strategy Journal* 1 (3-4): 191–205.

Hymer, S. (1976) *International operations of national firms: A study of foreign direct investment*. MIT Press, Boston.

Jarillo, J. C., and Martínez, J. (1991) *Estrategia Internacional. Más allá de la exportación*. McGraw Hill, Madrid.

Johanson, J., and Vahlne, J. E. (1977) The internationalization process of the firm—A model of knowledge development and increasing foreign market commitment. *Journal of International Business Studies* 8 (1): 23–32.

Johanson, J., and Vahlne, J. E. (1990) The mechanism of internationalization. *International Marketing Review* 7 (4): 11–24.

Johanson, J., and Vahlne, J. E. (2003) Business relationship learning and commitment in the internationalization process. *Journal of International Entrepreneurship* 1: 83–101.

Johanson, J., and Vahlne, J. E. (2009) The Uppsala internationalization process model revisited: From liability of foreignness to liability of outsidership. *Journal of International Business Studies* 40: 1411–1431.

Kalinic, I., and Forza, C. (2011) Rapid internationalization of traditional SMEs: Between gradualist models and born globals. *International Business Review*, 21 (4): 694–707.

Kamp, B. (2003) *Formation and evolution of international business networks": Kaleidoscopic organization sets*. Nijmegen, The Netherlands: Wolf Legal Publishers.

King, S. (2011) The Southern silk road. Turbocharging "South-South" economic growth. *Global Economics*, June, HSBC Global Research, London. Online at: http://www.research.hsbc.com/midas/Res/RDV?p=pdf&key=WZnyWSIf38&n=299714.PDF

Knight, G., and Cavusgil, S. (1996) The born global firm: A challenge to traditional internationalization theory. *Advances in International Marketing*, JAI Press, 11–26.

Kondratiev, N. (2002) The emergence of a New Techno-Economic Paradigm: the age of information and communication Technology (ICT), As time goes by: from the industrial revolutions to the information revolution. Oxford Scholarship.

Lafay, G., and Herzog, C. (1989) Commerce international: La fin des avantages acquis, económica, Pm's, Diffusion, Documentation Française.

Langhorne, R. (2001) *The coming of globalization: Its evolution and contemporary consequences.* Palgrave, New York

Luzarraga Monasterio, J. M. (2008) *Mondragon multi-location strategy—Innovating a human centred globalisation.* Mondragon University, Oñati, Spain.

Madsen, T. K., and Servaos, P. S. (1997) The internationalization of born globals: An evolutionary process. *International Business Review* 6 (6): 561–583.

McKinsey & Co. (1993) *Emerging exporters: Australia's high value-added manufacturing exporters.* Australian Manufacturing Council, Melbourne:.

Mediavilla, M., and Errasti, A. (2010) Framework for assessing the current strategic plant role and deploying a roadmap for its upgrading. An empirical study within a global operations network, *Proceedings of APMS 2010 Conference, Cuomo, Italy.*

Ohmae, K. (1987) *Beyond national boundaries: Reflections of Japan and the world.* Dow Jones-Irwin, New York.

Oviatt, B. M., and McDougall, P. P. (1994) Toward a theory of international new ventures. *Journal of International Business Studies* 25 (1): 45–64.

Oviatt, B. M., and McDougall, P. P. (2005) Defining international entrepreneurship and modelling the speed of internationalization. *Entrepreneurship Theory & Practice* 29 (5): 537–553.

Rialp, A., Rialp, J., Urbano, D., and Vaillant, Y. (2005b) The born-global phenomenon: A comparative case study research. *Journal of International Entrepreneurship* 3 (2): 133–171.

Sarasvathy, S. D. (2008) *Effectuation: Elements of entrepreneurial expertise.* Edward Elgar Publishing, Cheltenham, U.K.

Sen, A. (2002) How to judge globalism. American Prospect, Vol. 13.

Szabó, G. G. (2002) New institutional economics and agricultural co-operatives: A Hungarian case study. In *Local society and global economy: The role of co-operatives*, eds. S. Karafolas, R. Spear, and Y. Stryjan (pp. 357–378). Naoussa: Editions Hellin, ICA International Research Conference.

The Economist (2007) Globalisation's offspring.

Thompson, A., and Strickland, A. J. (2004) *Strategic management: Concepts and cases.* McGraw-Hill Irwin, New York.

UNCTAD (2011) World Investment Report 2011. *Non-equity modes of international production and development.* United Nations, New York and Geneva. Online at: http://www.unctad-docs.org/files/UNCTAD-WIR2011-Full-en.pdf

Vernon, R. (1966) International investment and international trade in the product life cycle, *Quarterly Journal of Economics* LXXX: 190–207.

Wilson, D., Trivedi, K., Carlson, S., and Ursúa, J. (2011) *The BRICs 10 years on: Halfway through the great transformation.* Global Economics Paper No. 208, Goldman Sachs Global Economics, Commodities and Strategy Research, New York. Online at: https://www.google.es/search?aq=0&oq=brics+10+&ix=seb&sourceid=chrome&ie=UTF-8&q=the+brics+10+years+on+halfway+through+the+great+transformation

Yip, G. S., and Bink, A. J. M. (2007) *Managing global customers: An integrated approach.* Oxford University Press, New York.

Zaheer, S. (1995) Overcoming the liability of foreignness. *Academy of Management Journal* 38 (2): 341–363.

chapter 2

GlobOpe Framework

Sandra Martínez, Miguel Mediavilla, and Ander Errasti

> *Being prepared means much, being able to wait means more, but to make use of the right moment means everything.*
>
> **Arthur Schnitzler**

> *If you want to succeed, don't stand looking at the stairs. Start climbing, step by step, until you reach the top.*

Contents

Introduction

In this chapter, we discuss:

- Stages to be considered in the internationalization of operations process in a company
- Factors to be considered in each stage
- The need for a reconfigurable production and logistic system
- GlobOpe framework

Internationalization Stages from the Early Beginning

The GlobOpe framework is centered in the more advanced stage of a company in terms of becoming global. Nevertheless, the maturity level of a company in terms of business and operations internationalization could be lower. Previously (see Chapter 1), the classification has been made for global, local, offshorer, and exporter companies. Companies are not tied to one of these strategies during their entire life cycle. Thus, they could change and adapt them. Following this strategy, three different behaviors and paths (Luzarraga, 2008) can be followed to achieve a rating of global competitor (Figure 2.1).

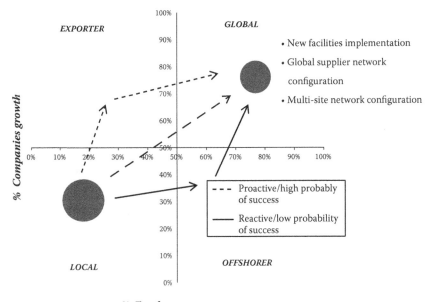

Figure 2.1 Paths in the internationalization strategy.

These roadmaps do not have the same probability of success or the same cost.

- The dotted line responds to a gradual itinerary and proactive based-on-sales growth with a high probability of success. According to Johansson and Vahlne (1977), the process of becoming an exporting firm can be divided into:
  - Sporadic export
  - Export with independent agents and/or export with own sales offices
- The discontinuous line usually responds to an internationalization process based on the acquisition of a foreign company with access to market share. It should be more radical, reactive, and expensive. Some companies acquire or become involved with existing companies by creating a shared production unit (this is called Joint Venture).
- The continuous line corresponds to a reactive strategy, initially increasing productive competition. It is unusual to find companies that have achieved a global position as a result of this strategy.

Local, exporting, or offshorer companies become global companies and they end up having some implementation of production units.

Each of these stages requires a change in the configuration of plants and warehouses as well as company organization. Thus, the author and contributors of this book encourage deep strategic thinking about the future production and supply network configuration using the GlobOpe framework in order to gain a competitive advantage.

Sporadic Exporting/Importing; Exporting/ Importing via Independent Agents; Commercial/Purchase Own Offices

The relocation strategy is to follow steps marked, always thinking about low wages and high reduction of other costs. At the initial step is to open a purchase office/commercial, which allows a company to get to know the local suppliers in the country chosen. This is followed for the creation of a logistics platform and, finally, purchasing an existing plant or building a new one according to local circumstances.

Import

When purchasing in low-cost countries, the logistics route complexity and supplier's management issues related to quality and financial points out the important barriers that may be encountered.

That is the reason why some enterprises start up this process with an agent or import trader. The agents and import traders aid in managing the custom and transport logistic issues. Nevertheless, the need to manage other key aspects, such as quality management, has made some normal traders into integral traders.

Export

One of the main entry barriers in new markets is the product distribution issue. Developing one's own distribution network takes years as well as a large investment. Thus, companies seek alternative ways to reduce the time to market and avoid entry barriers, especially in multi-customer environments.

Waters (2003) established different Western consumer routes to market models when investigating the Asian markets. These practices take into account the difficulties, cost, and time needed to develop one's own retail network starting from zero. Thus, the following models are proposed to penetrate into new markets (Table 2.1).

Table 2.1 Distribution models when facing a distribution in a new country

Models	Trade Sales by	Comments
Own retail		Rare and not usual
Joint venture		Appropriate for strong and volume retail formats
Own logistic resources to serve retail	Brand owner	Appropriate for dominant brands
Full agency distributor	Distributor	Appropriate for initial market entry. Not always appropriate for brand owners
Direct export to key account retailers and rest via local distributor	Brand owner for key accounts	
100% via local distributor	Mainly by distributor	For lower volume brands
Direct marketing	Agents	
Wholesalers	Variable	For major brands, used to serve smaller accounts via main distributor and its warehousing and transportation facility

Source: Waters, D. (2003) *Global logistics and distribution planning strategies for management.* Kogan Page, London.

Asia covers a vast area, with more than 60% of the world's population and around a quarter of its trade. The dominant area for the Asian economy is the southeast, particularly those countries in ASEAN (Association of South East Asian Nations). These include the Tiger economies, which include some of Hong Kong, Indonesia, Malaysia, Singapore, South Korea, Taiwan, and Thailand. Watching this picture, brand owners wishing to enter the market through retailers in these countries will have few doors on which to knock.

Things are very different in the less developed countries. There, it is essential for Western brands to make inroads into the traditional channels in order to succeed, even though this is very difficult. Underscoring this is the fact that there are few logistics companies offering Western-style services and capabilities outside Hong Kong and Singapore.

In the West, market entry for companies hinges upon access to the major retailers. In Southeast Asia, this is true only for Hong Kong, Singapore, and, to a lesser extent, Taiwan.

The route-to-market in most of Asia is heavily dependent upon assistance from third parties, with the first decision being whether to manufacture in the country or the region, or to import.

Brand owners from the West typically make three strategic errors, which account, in many instances, for a weak performance in China:

- Overestimating market size. To this day, one sees market forecasts based upon a top-down share of spending by the total population.
- Underestimating the sheer scale of expenditure on advertising and promotion needed, not merely to persuade consumers to buy, but to convert strongly rooted affinities to traditional goods. For example, the concept of mixing toasted cereals with plain cereals is conventional in the West, but a curiosity in China.
- Setting up factories in southern China where the more brand-receptive and trend-setting markets are far away in Shanghai and Beijing.

Another negative factor is protection of brand rights. A percentage of companies found their products being counterfeited, and a higher percentage of these were unable to find any remedy to the problem. One insight into such problems is that they partially reflect distribution inefficiencies. Local entrepreneurs "piggyback" the efforts of brand owners by operating a more effective route-to-market deployment.

Distribution, therefore, is often the most important factor—and challenge—for market success in China. Since China entered the World Trade Organization (WTO), the question of where to manufacture has at least become a financial rather than a regulatory issue. The plain fact is that no single distributor or distribution channel can sustain this kind of reach, for several reasons:

- China is, in effect, 24 different countries, based upon its provincial structure. Interprovincial transport is fraught with logistics and administration challenges, and the choice of freight mode is complex, including road and rail, and waterborne by inland or coastal vessel. Air transport is a limited option.
- Few major Western-style distribution companies operate across the major cities in China, and have on-tap service capacity for major new customers.
- Entry to the wholesaler network that dominates service to most outlets—and in turn a very large proportion of achievable volume— means establishing a large and complex patchwork of relationships generally segmented by geographical areas.

These difficulties spell out the need for overseas brand owners to have a very strong commitment to the consumer market in China. The majority of existing entrants have incurred losses for many years. This could not be tolerated unless they had a very optimistic view of the glittering prizes ahead in what (irrespective of regional recession today) may well prove to be the largest economy in the world before very long. Improved logistics will have to play a major part in this transformation (Waters, 2003).

Joint Production Facilities (Joint Venture)

Many times there are agreements between companies to facilitate the globalization effort. These agreements translate into a new company being formed, usually by two partners: the headquarters (wanting to globalize) and a company located in the foreign country. It may include one or more companies' suppliers also located in the foreign country.

Usually, the company's headquarters has strong know-how concerning the product and the manufacturing process; however it has to hurdle the barriers that are uneconomical for transporting and distributing the product from the country of origin.

The partner, that it is from destination country, is required, generally, sector expertise and it is responsible in the local adaptation of the product. It is normal to look for a partner that is a low-cost producer of a similar product in the region, in addition to having production capacity. The company benefits from the partner's expertise in product know-how, manufacturing process, and trade capacity of the headquarters.

These joint ventures are also crucial suppliers. The ideal supplier is one who has operations in a foreign country. The suppliers, who have operations that overlap with the external operations of the headquarters, are known as strategic suppliers and global suppliers because they supply not only the materials required to produce the good, but also the

Table 2.2 Advantages and disadvantages of a joint venture versus a company
with 100% Spanish capital

Advantages and Disadvantages of a Company with 100% Spanish Capital	
Advantages	Disadvantages
Total control	Lack of market and laws
Speed in decision making	Lack of local contacts
Protection of know-how	Higher risk to start business from scratch

Advantages and Disadvantages of a Joint Venture	
Advantages	Disadvantages
Market knowledge	Conflict management and internal competition
Partners have local contact	Problems and slowness in decision making
Supply network	Lack of control
Less risk and effort	Deferred interest
Ease in solving operational problems	Dependence of local partner
	Difficulty in communication and understanding with the partner

equipment used in production processes. **As part of a globalization project, suppliers often follow the headquarters' operations strategy in a foreign country.**

Increasingly, the supplier is in a position where it could make changes, especially in technology, product, or process. In the past few years, there has been an important development toward the extensive use of suppliers. It often takes over the process to provide important functional groups or subassemblies of the product. The selection of suppliers is based increasingly on the cost of the process.

Some authors have already identified the advantages or disadvantages of managing a 100% capital-owned company or joint venture.

New Production Facility Implementation

The Chief Operations Officers (COOs) usually try to implement a Joint Venture due to its advantages. Nevertheless, it is not always possible to find the suitable partners. In this case, the company can opt for the implementation of a production plant. This is one of the problems in which the GlobOpe model could be valuable.

According to Barnes (2002), this is the strongest step of any organization that seeks to internationalize. Besides the obvious economic investment, the establishment and subsequent management of the production facilities require a broad range of knowledge and skills related to operations management:

- As one must use as local staff an expatriate staff, it will require adequate capacity for human resource management.
- It should establish agreements with suppliers. The question is whether such supplies must be purchased onsite or be imported from known and trusted suppliers. Thus, it requires skill and experience in purchasing and supply management.
- It must set the logistics required to ensure that products are delivered to customers in the right way, which can force one to develop agreements with local logistics suppliers.
- It is possible to have to modify products and services to suit local requirements. This may involve having to establish design and development capabilities. Similarly, there may be a need for post-sales services for local customers.
- The company production can be exported; this makes the plant a part of the global supply network of the organization. In this case, it requires skills that are related to the managing of global operations.

Barnes (2002) also notes that some organizations that want to establish new production plants abroad **try to minimize the problems associated with the establishment through the acquisition of existing resources** in the country of destination. This means a scanning process of the resources through a process of merger and acquisitions (tasks, such as audits, due diligence, etc.). This may involve the purchase of assets (buildings, means of production, etc.) or the acquisition of part or even all existing commercial networks. In these cases, there might be a number of contingencies:

- The organization should be responsible for any agreements or arrangements that may have existed before, with staff employed, customers, suppliers, and others (e.g., agreements and commitments signed with local authorities), and possible legal rules that limit certain options. All of this forces the company to reconsider how advantageous the option is, particularly bearing in mind that it may be difficult to change in the short term those agreements and commitments.
- Another major challenge is its integration into the operating practices of the organization. It can cause problems due to contractual disputes between employees who are integrated and those in other parts of the organization. There also may be incompatibilities with the agreements with customers, suppliers, and other parts of the supply chain.
- It can cause problems as well due to differences in technology and production equipment purchasing practices and procedures.

- The potential incompatibility of the acquired information systems also can cause problems in information processing and communication between the acquired segment and the rest of the organization.

The required adjustments can be costly in time and resources.

The Need for a Reconfigurable Production and Logistic System

In the **global competitive playground**, the internationalization of operations has become a common trend among companies, mainly by multinationals, but **also by small to medium sized companies** (Corti, Egaña, and Errasti, 2008).

The main drivers or reasons that explain the internationalization phenomenon that was shown in Chapter 1 are offshoring, entering new markets, disaggregating the value chain, reengineering the value chain, and creating a new product or market (Ferdows, 1997; Farrell, 2006).

In a more and more dynamic and volatile market environment, Production Networks need to evolve over time and local approaches have to be integrated and coordinated with global infrastructures. With this emergence of the global supply and manufacturing sources, as well as the global market, such an operations network design will increasingly have to cover multiple regions and cope with a higher network complexity (e.g., a global supplier network).

In this context, the production and logistic system strategy or operations strategy conditions the decisions and reengineering projects to be carried out in a company in the medium and short term to improve the competitive advantage of the production and logistic chain. Operations strategy has to gain more effectiveness and efficiency over operations resources through defining and implementing suitable operations strategy decisions, managing the tangible resources, and developing operations capabilities in order to reach the performance objectives of the market requirements.

Nevertheless, market requirements are not static, but dynamic. Thus, the new paradigm in the context of Global Production Network design is that, if a company should adapt its operations to be as efficient and effective as possible, then there must be continuous reconfiguration of the Production Network and the new proposal should consider having the ability to be modified in the near future.

That means that the initial operations design and configuration also should consider the following properties on a network level (i.e., not only focusing on an isolated facility or plant, but with a network perspective):

- **Adaptability to a volatile and dynamic environment:** To balance long-term investments and resources needs with related risks depending on different scenarios.
- **Adaptability to make product demand changes:** To handle a variety of requirements that could change, such as product volumes. Thus, the proposed design should have the ability to be scalable and adjustable at reasonable costs to future needs.
- **Flexibility to product demand variety:** To handle a variety of requirements that could change the product mix. Thus, the proposed design should have the capacity to accomplish constraints due to the increase of information and material flow complexity related to product mix increase or decrease.
- **Upgradability to process engineering:** To select the most adequate equipment systems and technology suited to the environment where the facility operates and guarantees the quality process requirements.
- **Selective operability:** To select the markets and product segments that each facility should supply, assessing redundancy or uniqueness needs among facilities, and value propositions around service policy that consider responsiveness as a competitive advantage.
- **Contingency operability:** To select the design aspects and possible contingency plans to be put into practice to successfully confront the unforeseen events due to their high impact even if the probability of occurrence of this event is low.

All of these factors are summarized in the following characteristics (Mehrabi, Ulsoy, and Koren, 2000; Holweg and Pil, 2004):

- **Responsiveness:** process, product, and volume
- **Scalability**
- **Rapid adjustment of existing system**
- **Cost-efficient production**

As we mentioned above, in the context of Global Operations Network Strategy, companies need to reconfigure their Production Network and for this reason the new proposal should incorporate the ability to be modified in the near future.

This approach is, in fact, a changing and adaptive exercise for the company, therefore, the operations strategy and—by extension—**the international production and logistic network design should integrate dynamic capabilities evaluation** (Sweeney, Cousens, and Szwejczewski, 2007), as happens in any rationalization or restructuring of an operations network (Figure 2.2).

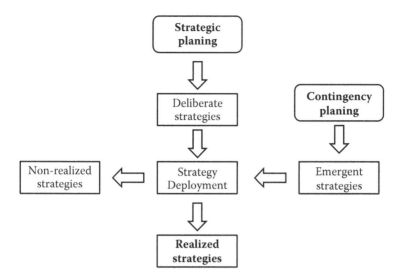

Figure 2.2 Deliberate and emergent strategies. (Adapted from Mintzberg, H., et al. (1996) *The strategy process,* 4th ed. Prentice-Hall, Hemel Hempstead, U.K.)

These dynamic capabilities are defined by Teece, Pisano, and Shuen (1997) as "the ability to achieve new forms of competitive advantage to emphasize two key aspects that were not the main focus of attention in previous strategy perspectives. The term *dynamic* **refers to the capacity to renew competences** to achieve congruence with the changing business environment. The term *capabilities* emphasizes the key role of strategic management in appropriately **adapting, integrating, and reconfiguring** internal and external organizational **skills, resources, and functional** competences to match the requirements of a changing environment."

GlobOpe Framework

Companies face both a variety of choices because of rapidly changing product and process technologies, and a variety of challenges because of a global competitive environment, informed customers, demanding owners, and environmental and political factors. Frameworks or models are helpful because they usually organize the important issues into a structure that enables companies to understand and be aware of them.

Although there are many researches about Global Operations, practical experience has shown that strategy-specific checklists are needed, which might raise awareness of the real success factors of the pursued goal. Such checklists could serve managers as experience-based guidelines to identify the most important criteria and thus avoid unpleasant surprises (Kinkel and Maloca, 2009).

Moreover, Vereecke and Van Dierdonck (2002) as well as Shi (2003) state that operations and supply chain management researchers should pay attention to providing understandable models or frameworks of international manufacturing systems that help managers design and manage their networks. Additionally, Avedo and Almeida (2011) also expose the need to build new conceptual frameworks that take into account the necessary requirements for the next generation of factories, which have to be modular, scalable, flexible, open, agile, and able to adapt, in real time, to the continuously changing market demands, technology options, and regulations.

The **GlobOpe model** is a **framework for the design and configuration process of a global production and logistic network that can be a useful management tool for small and medium enterprises (SMEs) and strategic business units (SBUs) steering committees** responsible for the effectiveness and efficiency of global operations.

Therefore, the two following questions that arise are:

1. Why is the GlobOpe model focused on SMEs and SBU?
2. How does one propose a model or framework that serves as a useful guideline for industrial SMEs and SBUs managers?

Concerning the first question, first of all, it is necessary to define what is understood as SMEs and SBU.

- According the European Commission, **SMEs** are companies with:
  - Number of employees < 250
  - Turnover < 50 M€
  - Balance sheet total < 43 M€
- A **Strategic Business Unit** is "a set of activities or a homogeneous business from a strategic point of view, for which it is possible to formulate a common strategy and in turn different from the strategy for other activities and/or strategic units. Thus, the strategy of each unit is autonomous, but not independent of other units because all of them are integrated in the company strategy. Then, the company can be considered as a set of several SBUs, each of them offering profit opportunities and different growth and/or requiring a dissimilar competitive approach" (Menguzzato and Renau, 1991).

SMEs' relevance is increased in the worldwide economic system (Knight, 2001; OECD, 1997; Shrader, Oviatt, and McDougall, 2000). According to EUROSTAT, in the sole European Union in 2008, there were 20.7 million SMEs. They account for 99.8% of the total number of enterprises and employ 67.4% of the population. Moreover, a recent study of the European Commission (2010) reports that "internationally active SMEs report an employment growth of 7% versus only 1% for SMEs without any international activities" and "26% of internationally active SMEs introduced products or services that were new for their sector in their country; for other SMEs, this is only 8%."

Due to the rapid changes that have occurred during the last decades, almost every company or enterprise is affected by at least some kind of international challenge. Nevertheless, in the case of SMEs, they have to cope with more difficulties because of having limited resources, limited market knowledge, limited use of networks, and limited international experience of the entrepreneurs (Kalinic and Forza, 2011), and these same complications can affect SBUs as well. Consequently, the SMEs' and SBUs' internationalization process merits great attention.

Regarding the question #2, the model intends to fill the gap left by the production systems (i.e., Toyota PS, Volvo PS, Bosch Siemens PS, etc.) and the Lean manufacturing programs. The Toyota PS and Lean techniques lead the way to excellence in a stable environment, but they are not suitable in a dynamic market environment where new facility implementation, supplier network development, and reconfiguration of an existing network are needed (Mediavilla and Errasti, 2010). In these cases, **effectiveness rather than efficiency is the primary goal in the first stage of implementation, and more value chain activities must be taken into account when reconfiguring the network**.

The network analysis and design process is based on another approach called KATAIA (Errasti, 2006). The KATAIA mode takes into account that, given the implications of configuring operations such as new facility implementation, global supplier network development, and multiplant network reconfiguration, deciding whether and how to configure operations should be considered a strategic issue for the company, and thus the concepts identified in the literature were set around the steps that are typically needed in a strategy development process.

Authors such as Acur and Biticci (2000), state that for a dynamic strategy development process five stages (inputs, analysis, strategy formulation, strategy implementation, and strategy review) are needed and that management and analytical tools can be used for this purpose.

We have adopted this approach; nevertheless, the GlobOpe framework simplifies Acur and Biticci's method and adapts it to the operational strategy business units, taking the following factors into consideration:

The methodology/guide takes into account the position of the business unit in the value chain (Porter, 1985) and sets the stage that should help value creation. An analysis stage is used to analyze the factors (Anumba, Siemieniuch, and Sinclair, 2000; Boddy and Macbeth, 2000; Hobbs and Andersen, 2001; Acur and Bitticci, 2000) and choose the content of the strategy (Gunn, 1987). The analysis contributes to a definition or formulation of the new facility ramp up process, and then a deployment stage of the formulated design is set (Feurer, Chaharbaghi, and Wargin, 1995). The deployment

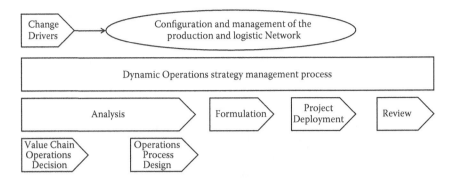

Figure 2.3 Schematic representation of the GlobOpe framework.

is a project-oriented task (Marucheck, Pannesi, and Anderson, 1990), where a process of monitoring and reviewing to facilitate the alignment of the organization to the operations strategy is set (Kaplan and Norton, 2001).

All of these properties should contribute to the need to propose a **reconfigurable Production and Logistic Network.**

A scheme of the framework is illustrated in Figure 2.3.

The key **operations strategy** decisions to be made regarding global production and logistic network configuration and design in the internationalization process include:

1. Supply sources location (own and not own)
2. Facilities, suppliers, and warehouse strategic role
3. Integration or fragmentation of productive and logistic operations: Make or buy decisions
4. Service delivery strategy: Supply strategy/manufacturing strategy/ purchasing strategy
5. Global Operations Network: Distribution network/manufacturing network/suppliers network

Nevertheless, the author of and contributors to this book consider that there are three main problems related to operations configuration where the above decisions should be reviewed:

- New facilities implementation
- Global supplier network configuration
- Multisite network configuration

Figure 2.4 shows a scheme of the complete GlobOpe framework, which considers the three problems mentioned above.

When companies cope with the situation of opening a new facility or when they need to develop a supplier network, there are two key processes

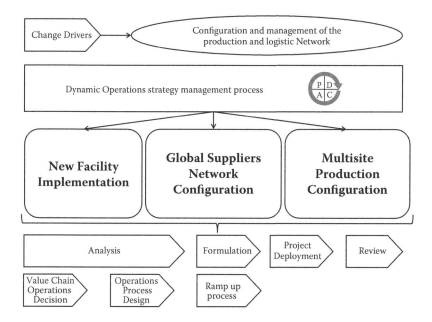

Figure 2.4 Schematic representation of the complete GlobOpe framework.

of the value chain (Porter, 1985) that should be reviewed in depth because of their impact in the operations performance: **new product development (NPD) and order fulfillment (OF).**

NPD is the term used to describe the complete process of bringing a new product to market. A product is a set of benefits offered for exchange and can be tangible (i.e., something physical one can touch) or intangible (such as a service, experience, or belief). There are two parallel paths involved in the NPD process: (1) idea generation, product design, and detail engineering and (2) market research and marketing analysis. Companies typically see new product development as the first stage in generating and commercializing new products within the overall strategic process of product life-cycle management used to maintain or grow their market share.

Order fulfillment is the complete process from point of sales inquiry to delivery of a product to the customer.

Figure 2.5 shows the GlobOpe Model focused on these two key process analyses.

GlobOpe for New Facility Implementation

New manufacturing and supply configurations, which companies who are undergoing the internationalization process in order to enter new markets must face when installing new facilities overseas, is a topic that

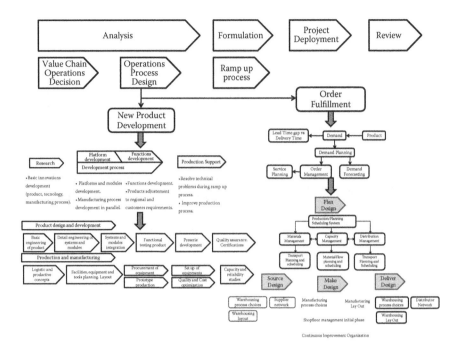

Figure 2.5 GlobOpe framework focused on two key processes: NPD and OF.

is becoming relevant in the science of operations management (Ferdows, 1997; Errasti, 2011).

Additionally, the coordination between agents involved in the supply network chain and the supply strategy response to a highly dynamic and volatile environment when entering a new market could cause ramp up delays in time and volume, especially when new factors are introduced, which then leads to production losses (Abele et al., 2008).

When assessing production locations abroad, companies tend to underestimate the necessary ramp up times for securing process reliability, quality. and productivity. The ramp up concept describes the period characterized by product and process experimentation and improvements (Terwisch and Bohn, 2001), which, strictly speaking, starts with the first unit produced and ends when the planned production volume is reached (T-Systems, 2010). Nevertheless, in order to manage such a ramp up with a high degree of precision, an initial planning period phase is necessary, starting with the design of the product, the process, and the supply chain network (Kurttila, Shaw, and Helo, 2010), (Sheffi, 2006).

A study carried out in 39 internationally active German companies showed that not only the small firms, but also the larger companies tended to heavily underestimate ramp up times and coordination costs for foreign production sites. Specifically, on average, ramp up times were

2.5 times longer than originally planned. The absolute time required for the ramp up of overseas production sites until production processes run smoothly has established ranges, in almost all cases, of two to three years. Ramp up times do not only entail higher coordination costs, they also can considerably affect the calculated amortization time, which for many companies is the decisive criterion that tips the scale for or against an off-shoring engagement (Kinkel and Maloca, 2009).

As can be seen, a typical network reorganization will require many projects, each involving transfers of products between plants. This may entail ramping up or down existing plants, developing new sites, or closing existing ones. In some companies, such activities are managed by experienced project managers using ad hoc, intuitive processes. However, given the scale of change and the interlinkages between the various projects, it may be beneficial to build systematized expertise in these processes so that projects proceed more smoothly and with a higher probability of successful completion on time (Christodoulou et al., 2007). Hence, the author and contributors of this book consider that the GlobOpe Model for New Facilities Implementation could be used as a systematic guideline to face these problems.

Bearing in mind the facility life cycle, this model proposes potential methods and techniques for aiding the decision process (Figure 2.6).

This model could serve as a management tool for a steering committee when they have to deal with some of the following cases as shown in Table 2.3 and Table 2.4.

GlobOpe for Global Supplier Network Development

There are companies that have expanded their operations throughout the world. As an example, there are Japanese companies that do R&D in networks involving American and Asian researchers, do detailed engineering in Bangalore (India), produce components in Taiwan, and carry out quality control and assembly in Tianjiin (China) in order to sell the final product in Europe and America. The implications for purchasing professionals are that their operations must be professionalized, streamlined, and globalized due to this new context (Van Weele and Rozemeijer, 1996). Even in the current competitive markets, companies are turning to hybrid purchasing organizations in order to leverage global sourcing benefits (Trautmann, Bals, and Harmann, 2009).

Some authors (Leenders et al., 2002) consider that global purchasing management is one of the first steps in defining and designing a global supply chain development.

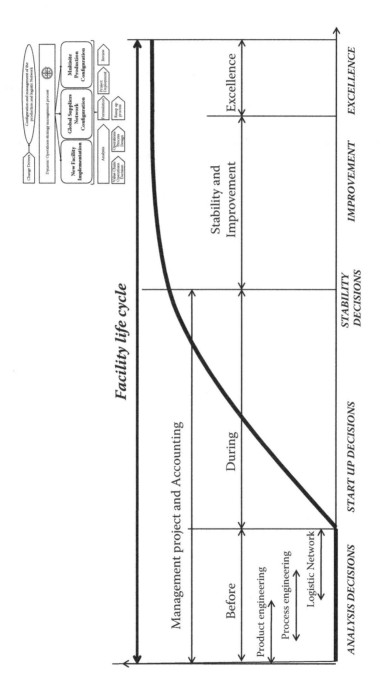

Figure 2.6 Scheme of the GlobOpe Model for New Facilities Implementation.

Table 2.3 Cases where the GlobOpe Model is useful for New Facilities Implementation

TO BE	New Facility in a New Geographic Area and/or Migrating from exporter to Global		
AS IS	One Facility	Several Domestic Facilities	Several Multisited Facilities
1 Business Unit	✓	✓	✓
2 Business Units	✓	✓	✓

Table 2.4 Cases where the GlobOpe Model is useful for New Facilities Implementation

	Supplier Network Design	
	Server, Outpost, or Offshore role	Contributor, Lead, or Source role
New facility	**GlobOpe Model for new facilities implementation**	GlobOpe Model for global suppliers network configuration
Multiplant/ monoproduct	GlobOpe Model for global suppliers network configuration	GlobOpe Model for global suppliers network configuration
Multiplant/ multiproduct	GlobOpe Model for global suppliers network configuration	GlobOpe Model for global suppliers network configuration

Global or international purchasing is defined as *the activity for search-ing and obtaining goods, services, and other resources on a possible worldwide scale, to comply with the needs of the company and with a view to continuing and enhancing the current competitive position of the company* (Van Weele, 2005).

Moreover, global purchasing can be the result of a reactive, opportu-nistic decision to decrease the purchasing cost of one item, but it also can be a strategic and coordinated effort to proactively enhance the competi-tive position of the company. It includes all phases of the purchasing pro-cess, from before the definition of the specification list, through supplier selection and buying, to the follow-up and evaluation phase.

Trent and Monczka (2002) propose five stages in the purchasing inter-nationalization process, as seen in Table 2.5.

Moreover, according to Trent and Monczka (2003), an important pre-requisite for implementing global sourcing is the criticality of aligning global sourcing strategy with organizational design.

Table 2.5 Stages in the purchasing internationalization process

Stage 1	Engage in domestic purchasing only
Stage 2	Engage in international purchasing as needed
Stage 3	International purchasing as part of sourcing strategy
Stage 4	Integration and coordination of global sourcing strategies across worldwide buying locations
Stage 5	Integration and coordination of global sourcing strategies with other functional groups

Source: Adapted from Trent, R. J., and Moneczka, R. M. (2002) Pursuing competitive advantage through integrated global sourcing. *Academy of Management Executive* 16 (2): 66–80.

Keeping in mind the purchasing stages proposed by Trent and Monczka (2002), we define and summarize three stages: (1) when companies only purchase in domestic or local markets, (2) when the companies engage in a close international purchasing strategy, that is to say, with nearby countries, and (3) when they develop a worldwide purchasing strategy (Figure 2.7).

The GlobOpe Model for Global Suppliers Network Configuration could be used as an assessment tool for a steering committee to help them in the global supplier development. This model proposes purchasing strategies, policies, levers, and techniques for aiding the decision process and identifying the next steps that the company needs to go through in the next stage of the purchasing internationalization process.

This model could be a management tool for a steering committee when they have to deal with some of the following cases (Table 2.6):

GlobOpe for Multiplant Configuration

As stated in Shi and Gregory (1998), the multidomestic approach to network management that is characterized by a weak coordination that involves the development of more or less autonomous manufacturing units geographically located close to target markets is no longer enough to succeed in a global environment. During the last few decades "early movers" or multinational corporations have attempted to globalize their geographically dispersed plants by coordinating them with a synergetic production network.

What is needed is a globalized approach for network design and management that involves closely managed coordination of a dispersed manufacturing system and integration of both product design and development and production. The manufacturing system is seen as a unified whole with a mechanism for sharing knowledge, with elements of the task being performed in the most advantageous areas. An international manufacturing system thus may be seen as a factory network with matrix connections, in contrast to the linear system of a factory. The production plants are part of a

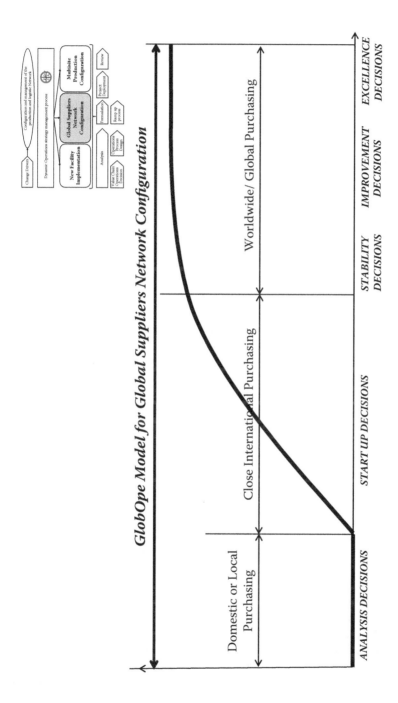

Figure 2.7 Scheme of the GlobOpe Model for Global Suppliers Network Configuration.

Table 2.6 Cases where the GlobOpe Model for Global Suppliers Network Configurations is useful

	Supplier Network Design	
	Server, Outpost, or Offshore Role	Contributor, Lead, or Source Role
New facility	GlobOpe Model for new facilities implementation	**GlobOpe Model for global suppliers network configuration**
Multiplant/ monoproduct	**GlobOpe Model for global suppliers network configuration**	**GlobOpe Model for global suppliers network configuration**
Multiplant/ multiproduct	**GlobOpe Model for global suppliers network configuration**	**GlobOpe Model for global suppliers network configuration**

manufacturing network and assume that whenever part of a network changes, it is likely to have implications for the entire network.

The question that arises is **how to deploy the Operations Strategy in a multilocation Global Operations Network** (GON), i.e., how to balance the different competences and responsibilities throughout the different factories or facilities, taking into account that the different units of the GON could assume different strategic responsibilities for themselves or for the whole GON.

This approach requires analyzing, defining, and upgrading the strategic role of manufacturing and production facilities. We take for this purpose the possible manufacturing roles, taking into account **the site competence** and **the strategic reason for establishing** and **exploiting the plant** (Ferdows, 1989). These definitions could be useful for the description and assessment of today's network of plants. Ferdows states that there are more developed manufacturing roles (lead, source, contributor) and less sophisticated roles (offshore, server) with less value chain activity (purchasing, process engineering, etc.) autonomy.

A company could start from one single plant for a regional or domestic market, and evolve to a multidomestic, uncoordinated network with isolated plants or lower role plants (offshore). Nevertheless, this network could be improved by seeking specializations in manufacturing roles. Some companies attempt to centralize some productions in specific or focused plants, with other distributed plants for their local markets (contributor) and take the most for overseas plants (source). In the most advanced stage (Figure 2.8), there is not only an information and materials flow, but there is even knowledge transfer in core value chain activities, such as new product development.

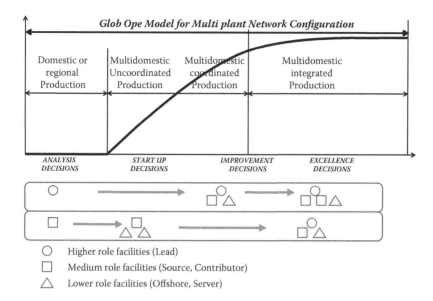

Figure 2.8 Scheme of the GlobOpe Model for Multiplant Network Configuration.

Practical experience has shown that strategy-specific methods for role upgrading are needed, which might raise awareness of the real success factors of the pursued goal. Such methods could serve managers as experience-based guidelines to identify the most important criteria and thus avoid unpleasant surprises (Kinkel and Maloca, 2009).

References

Abele, E., Meyer, T., Näher, U., Strube, G., and Sykes, R. (2008) Global production: A handbook for strategy and implementation. Springer, Heidelberg, Germany.

Acur, N. and Biticci, U. (2000) Active assessment of strategy performance in proceedings of the IFP WG 5.7. Paper presented at the International Conference on Production Management, Tromso, Norway, June 28–30.

Anumba, C. J., Siemieniuch, C. E., and Sinclair, M. A. (2000) Supply chain implications of concurrent engineering. *International Journal of Physical Distribution and Logistics* 30 (7/8): 566–597.

Azvedo, A., and Almeida, A. (2011) Factory templates for digital factories framework. Robotics and Computer-Integrated Manufacturing 27: 755–771.

Barnes. D. (2002) The complexities of the manufacturing strategy formation process in practice. *International Journal of Operations & Production Management* 22 (10): 1090–1111.

Boddy, D., and Macbeth, D. (2000) Prescriptions for managing change: A survey of their effects in projects to implement collaborative working between organizations. *International Journal of Project Management* 18: 297–306.

Christodoulou, P., Fleet´s D., Hanson, P., Phaal, R., Probert, D., and Shi, Y. (2007) *Making the right things in the right places. A structured approach to developing and exploiting "manufacturing footprint" strategy*, IFM, University of Cambridge, U.K.

Corti, D., Egaña, M. M., and Errasti, A. (2008) Challenges for off-shored operations: Findings from a comparative multi-case study analysis of Italian and Spanish companies. Paper presented at the EurOMA Congress, Groningen, The Netherlands, June 15–18.

Errasti, A. (2006) KATAIA. Modelo para el análisis y despliegue de la estrategia logística y productiva. PhD diss., Tecnun (University of Navarra), San Sebastian, Spain.

Errasti, A. (2011) *International manufacturing networks: Global operations design and management*. Servicio Central de Publicaciones del Gobierno Vasco, San Sebastian, Spain.

European Commission. (2010) Annual report on the European Union's development and external assistance policies and their implementation in 2009.

Farrell, D. (2006) *Offshoring. Understanding the emerging global labor market*. McKinsey Global Institute, Harvard Business School Press, Boston.

Ferdows, K. (1989) Mapping international factory networks. In K. Ferdows (ed.), Managing international manufacturing. New York: Elsevier Science Publishers. 3–21.

Ferdows, K. (1997) Making the most of foreign factories. *Harvard Business Review* March–April: 73–88.

Feurer, R., Chaharbaghi, K., and Wargin, J. (1995) Analysis of strategy formulation and implementation at Hewlett Packard. *Management Decision* 33 (10): 4–16.

Gunn, T. G. (1987) *Manufacturing for competitive advantage: Becoming a world class manufacturer.* Ballinger Publishing Company, Boston.

Hobbs, B., and Andersen, B. (2001) Different alliance relationships for project design and execution. *International Journal of Project Management* 19: 465–469.

Holweg, M., and Pil, F. (2004) *The second century: Moving beyond mass and lean production in the auto industry*. MIT Press, Cambridge, MA and London.

Johansson, J., and Vahlne, J. E (1977) The mechanism of Internationalisation. *International Marketing Review* 7.

Kalinic, I., and Forza, C. (2011) Rapid internationalization of traditional SMEs: Between gradualist models and born globals. *International Business Review*. 21 (4): 694–707.

Kaplan, R. S., and Norton, D. P. (2001) *The strategy focused organization*. Harvard Business School Press, Boston.

Kinkel, S., and Maloca, S. (2009) Drivers and antecedents of manufacturing offshoring and backshoring. A German perspective. *Journal of Purchasing & Supply Management* 15: 154–165.

Knight, G. A. (2001) Entrepreneurship and strategy in the international SME. *Journal of International Management* 7: 155–171.

Kurttila, P., Shaw, M., and Helo, P. (2010) Model factory concept-enabler for quick manufacturing capacity ramp up, European Wind Energy Conference and Exhibition, Warsaw, Poland.

Leenders, M., Fearon, H. E., Flynn, A. E., and Johnson, P. F. (2002) *Purchasing and supply management*. McGraw Hill/Irwin, New York.

Luzarraga, J. M. (2008) *Mondragon multilocation strategy: Innovating a human centred globalisation*. Mondragon University, Oñati (Spain).

Marucheck, A., Pannesi, R., and Anderson, C. (1990) An exploratory study of the manufacturing strategy in practice. *Journal of Operations Management* 9 (1): 101–23.

Mediavilla, M., and Errasti, A. (2010) *Framework for assessing the current strategic plant role and deploying a roadmap for its upgrading. An empirical study within a global operations network.* APMS, Cuomo, Italy.

Mehrabi, M., Ulsoy, A., and Koren, Y. (2000) Reconfigurable manufacturing systems: Key to future manufacturing. *Journal of Intelligent Manufacturing* 11 (4): 403–419.

Menguzzato, M., and Renau, J. J. (1991) *La dirección estratégica de la empresa.* Ed. Ariel, Barcelona.

Mintzberg, H., Lampel, J., Quinn, J. B., and Ghoshal, S. (1996) *The strategy process,* 4th ed. Prentice-Hall, Hemel Hempstead, U.K.

OECD. (1997) The OECD report on regulatory reform: Synthesis. Paris: Organisation for Economic Co-operation and Development.

Porter, M. (1985) *Competitive advantage: Creating and sustaining superior performance.* Free Press, New York.

Sheffi, Y. (2006) *La empresa robusta*, Lidl, Madrid.

Shi, Y. (2003) Internationalization and evolution of manufacturing systems: Classic process models, new industrial issues, and academic challenges. *Integrated Manufacturing Systems* 14: 385–396.

Shi, Y. and Gregory, M. (1998) International manufacturing networks —to develop global competitive capabilities. *Journal of Operations Management* 16: 195–214.

Shrader, R. C., Oviatt, B. M., and McDougall, P. P. (2000) How new ventures exploit trade-offs among internationization of the 21st century. *Academy of Management Journal* 43 (6): 1227–1247.

Sweeney, M., Cousens, A., and Szwejczewski, M. (2007) International manufacturing networks design: A proposed methodology. Paper presented at the EurOMA Conference, Ankara.

Teece, D. J., Pisano, G., and Shuen, A. (1997) Dynamic capabilities and strategic management. *Strategic Management Journal* 18 (7): 509–533.

Terwisch, C., and Bohn, R. (2001) Learning and process improvement during production ramp up. *International Journal of Production Economics* 70.

Trautman, G., Bals, L., and Harmann, E. (2009) Global sourcing in integrated network structures: The case of hybrid purchasing organizations. *Journal of International Management* 15: 194–208.

Trent, R. J., and Monczka, R. M. (2002) Pursuing competitive advantage through integrated global sourcing. *Academy of Management Executive* 16 (2): 66–80.

Trent, R .J., and Monczka, R. M. (2003) Understanding integrated global sourcing. *International Journal of Physical Distribution and Logistics Management* 33 (7): 607–629.

T-Systems Enterprise Services Gmblt. (2010) *White paper ramp up management. Accomplishing full production volume in-time, in-quality and in-cost.* Global Business Development and Consulting.

Van Weele, A. J. (2005) *Purchasing and supply chain management.* Thompson Learning, London.

Van Weele, A. J., and Rozemeijer, F. A. (1996) Revolution in purchasing: Building competitive power through proactive purchasing. *European Journal of Purchasing and Supply Management* 2: 153–160.

Vereecke, A., and Van Dierdonck, R. (2002) The strategic role of the plant: Testing Ferdow's model. *International Journal of Operations and Production Management* 22: 492–514.

Waters, D. (2003) *Global logistics and distribution planning: Strategies for management.* Ediciones Kogan Page, London.

chapter 3

Operations Strategy and Deployment

Kepa Mendibil, Martin Rudberg, Tim Baines, and Ander Errasti

> *The formulation of a problem is often more essential than its solution*
>
> **Albert Einstein**

Contents

Introduction

In this chapter, we discuss:

- Competitive advantage and generic strategies for business
- Product-market placement
- Business strategy and operations strategy
- Performance objectives: Quality service
- Performance objectives: Operations costs
- Operations strategy and operations design and management
- Operations strategy and supply strategy
- Operations strategy: Manufacturing and productive sustainable growth models
- Manufacturing servitization

Competitive Advantage and Generic Strategies for Business

Competition in an industry is based on competitive forces. The state of competition in an industry depends on **five basic forces**: the industry's position among current competitors, the bargaining power of customers, the threat of new entrants, the threat of substitute products or services, and the bargaining power of suppliers (Porter, 1998).

In this context, **competitiveness** is the ability of a company in a competitive environment to increase market share or profitability in a sustainable way (Porter, 1985).

Strategy is understood as the creation of a unique and valuable position among competitors. Value is the money a customer is ready to pay. Porter (1985) distinguishes two generic strategies for business: **product leadership and cost leadership.** These are created through connecting basic activities in a chain or value chain (Figure 3.1).

Kaplan and Norton (2001) formulated another framework for a company's value chain, one which distinguishes between the large wave (innovation) and short wave (operations) in the creation of value (Figure 3.2).

In addition to Product and Cost leadership, Kaplan and Norton (2001) added a third business strategy called Customer Intimacy. Each of these

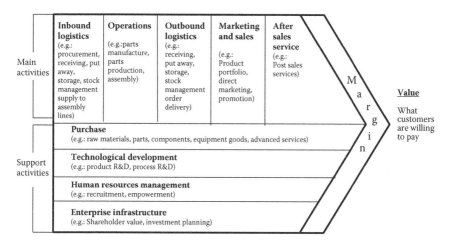

Figure 3.1 The generic value chain of a company. (Adapted from Porter, M. (1980) *On competition*. Harvard Business Review, New York.)

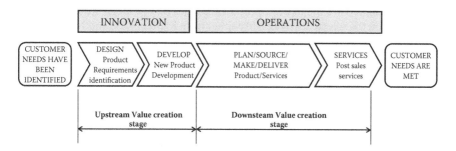

Figure 3.2 Large wave and short wave in value creation. (Adapted from Kaplan, R. S., and Norton, D. P. (2001) *The strategy focused organization: How balanced scorecard companies thrive in the new business environment.* Harvard Business School Press, Boston.)

generic strategies involves reaching a basic level of performance in certain processes and a qualitative jump up in other processes that reinforce the adopted strategy (Figure 3.3).

Product-Market Placement

In terms of the generic strategies for business, the company has to define the Product-Market strategy and the strategy for each business (when a company has more than one). Then, the company has to deploy for each product-market the **strategic positioning and placement in each** strategic **market segment.**

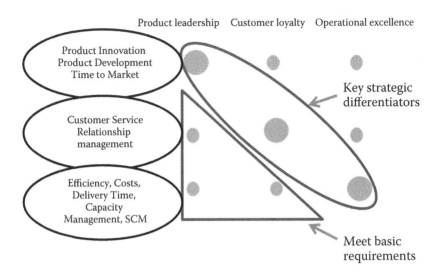

Figure 3.3 Shown are the most important activities (key strategic differentiators) depending on the selected strategy. (Adapted from Kaplan, R. S., and Norton, D. P. (2001) *The strategy focused organization: How balanced scorecard companies thrive in the new business environment.* Harvard Business School Press, Boston.)

Intel is a world leader in the development of processor technology for the IT sector. Intel's product range includes processors, motherboard chipsets, network controllers, and solid state drives. Intel's strength in the market is the result of a combination of advanced processor and chipset technology design with world class manufacturing capability. Their business strategy is focused on *product leadership* and this is enabled through robust research, innovation, product development, and manufacturing processes. Intel's technology and product roadmap includes future offerings that should enable the company to remain the market leader in microprocessor technology in years to come.

Since its origins back in the 19th century **IBM** has remained a leading global information technology organization. A key component in the success of IBM has been the ability to continuously adapt their business model in line with changing market conditions. For many years IBM was the world's major personal computer manufacturer but, as part of one the many re-structuring processes, this side of the business was sold in 2005. Currently, IBM is considered to be a systems integrator with a *customer intimacy* value proposition at the core of their strategy. Developing close relationships with customers in order to understand their needs and deliver bespoke technological solutions is a core competence for IBM. Its key customer base includes major multinational original equipment manufacturers, retailers, and service providers. This strategy is underpinned by their *product leadership* focus on information technology solutions. This combination of *customer intimacy* and *product leadership* strategies is evidenced by the fact that IBM is one the of the largest management consultancies in the world and holds more patents than any other U.S.-based technology firm.

The global airline industry was revolutionized with the emergence of the so-called low cost airlines. Up until that point the sector was characterized by airline companies focused on brand development and customer intimacy. The business models of companies like **Southwest Airlines, Easyjet,** and **Ryanair** were based on a *cost leadership* and *operational excellence* strategy, which at the time was new to the industry. These companies acquired a significant share of the market whilst maintaining high levels of profitability by focusing on the efficiency of operations, eliminating "bells and whistles" from their service, and increasing asset utilization ("time in the air").

IKEA is one of the leading home furnishing companies in the world. The company has reached annual sales of €23.1 billion (FY 2010) and has some 127,000 employees. IKEA has more than 620 million visitors per year in more than 300 stores all over the world. In addition to the visitors in the stores, some 712 million visitors are tracked entering the IKEA website. IKEA's main marketing channel is its catalog, which is distributed worldwide in 197 million copies (in 61 different editions and 29 different languages) displaying some of IKEA's more than 9500 selling items. IKEA's growth has been tremendous and sales are still growing. Currently IKEA plans to open 10–20 new stores every year with a goal to double sales every fifth year. Considering the pace of growth in sales, the many stores and warehouses, and the fact that some business areas change up to 30% of its assortment every year, IKEA's business and market strategies must be aligned with its operations strategy to stay competitive in the home furnishing market.

With its vision to "create a better everyday life for many people," IKEA focuses on low cost (cost leadership), but with reasonable quality, appealing designs, and adequate functionality. To keep up with the business and market strategy, and still delivering high profitability, the supply chain and logistics management need tight control and high levels of visibility to keep costs down and to avoid obsolete inventory and/or stock outs. Given the IKEA strategy, the main performance objectives are sales growth and production and logistics efficiency. Since IKEA does not own any manufacturing facilities, they are heavily dependent on the contribution of its entire network of its stores, its approximately 30 distribution centres, its 1400 suppliers, and its logistic partners. IKEA has therefore centralized supply chain and logistics planning where they also include suppliers' capacities in their planning process. Being a make-to-stock (MTS) producer, IKEA also focuses on stock reductions, yet without jeopardizing service levels and stock availability in the stores. Hence, there are many trade-offs that need to be reconciled in the regular meeting between the business/market and operations functions at IKEA.

To facilitate this analysis, companies can use management tools, such as:

- The BCG (Boston Consulting Group) matrix, which relates two variables: market growth (market attraction) and market share (competitive position and the ability to get cash) (Figure 3.4).
- Ansoff's strategic options and positioning matrix (Figure 3.5).

SME Dilemma

Single and medium enterprises (SMEs) are usually focused in a region/ nation and specific sectors where their markets are mostly domestic.

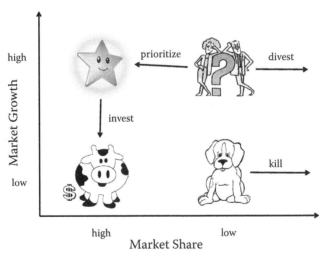

Figure 3.4 The Boston Consulting Group matrix.

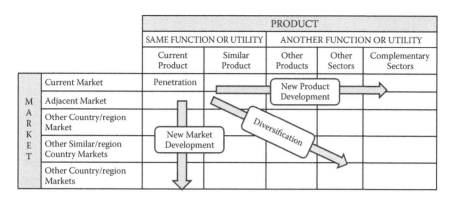

		PRODUCT				
		SAME FUNCTION OR UTILITY		ANOTHER FUNCTION OR UTILITY		
		Current Product	Similar Product	Other Products	Other Sectors	Complementary Sectors
M A R K E T	Current Market	Penetration		New Product Development		
	Adjacent Market					
	Other Country/region Market		New Market Development	Diversification		
	Other Similar/region Country Markets					
	Other Country/region Markets					

Figure 3.5 Ansoff matrix for strategic positioning by Ansoff. (Adapted from Sainz Vicuña, J. M. (2006) *El plan de marketing en la práctica*. ESIC, Madrid.)

Adopting a business strategy is a crucial decision. For example, **when adopting a Product Leadership** strategy, migrating to more value-added sectors (e.g., migration from being an automotive sector supplier to a supplier in the aeronautics and wind energy sectors), or a strategy of **Customer Intimacy** following the internationalization process of their customers or original equipment manufacturers (OEMs) (e.g., opening a new facility in China next to the main customer facility).

Linn Products Ltd. is a medium-sized organization that designs and manufactures audio equipment for the high-end market. Linn aims to achieve *product leadership* though developing products with the highest quality of sound reproduction. The company has developed a highly integrated research, development, and manufacturing process to achieve their strategy. Their manufacturing process is based on the "one stage built" philosophy, where technology and human skill are combined to ensure the quality and reliability of their products. Their supply chain is designed in a way that ensures that all parts and components complement the product leadership strategy.

Highland Spring Ltd. is a leading bottled water brand in the United Kingdom. A key building block of Highland Spring's success has been establishing a clear focus on brand leadership. This brand leadership strategy was underpinned by supporting their high quality product with strategic partnerships and high profile marketing campaigns and sponsorships. To support the brand's leadership value proposition, Highland Spring continuously strives for operational excellence through the development of highly flexible and efficient production and supply chain processes. The developed production and logistic infrastructure, in conjunction with continuously searching for operational excellence in their production system, has enabled the company's exponential growth.

Business Strategy and Operations Strategy

The production and logistics system strategy, that is, **the operations strategy, has to be aligned with the business's product/market strategy**, which conditions the decisions and reengineering projects to be accomplished in the medium and short term in a company in order to improve the competitive advantage of the supply chain (Figure 3.6).

In this context, Slack and Lewis (2002) define operations strategy as:

> The total pattern of decisions which shape the competences and capabilities of any type of operations (productive or logistics) and their contribution to the business strategy, through the contribution of operations resources to market requirements.

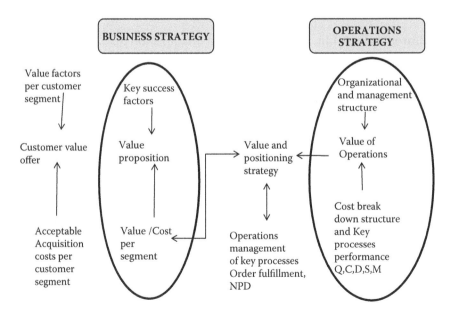

Figure 3.6 Business strategy and operations strategy connections. (Adapted from Monczka et al. (2009) *Purchasing and supply chain management.* Seng Lee Press, Singapore.)

The operations strategy has to improve effectiveness and efficiency over operations resources through defining and implementing the suitable operations strategy decisions, managing tangible resources, and developing operations capabilities to reach the performance objectives of the market requirements.

This has to do mostly with the location of the supply, production, and distribution nodes, and the physical design and organization of these nodes' processes. This problem arises when planning a manufacturing and logistics network or when restructuring an existing one (Vereecke and Van Dierdonck, 2002).

In the following section, the performance objectives (Figure 3.7) associated with market requirements will be explained.

Performance Objectives: Quality Service

The role of service is to provide "time and place utility" in the transfer of goods and services to customers (Christopher, 2005).

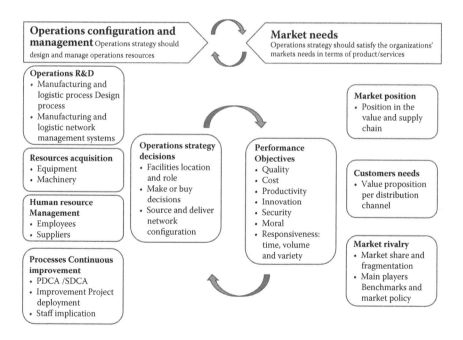

Figure 3.7 Operations strategy and the reconciliation of market requirements with operations resources. (Adapted from Slack, N., and Lewis, M. (2002) *Operation strategy,* 2nd ed. Prentice Hall, Upper Saddle River, NJ.)

Quality Service and the Components of Customer Service

La Londe and Zinszer, cited by Christopher (2005), suggested that customer service could be divided into the following elements: Pretransaction,
transaction, and posttransaction (Table 3.1).

Table 3.1 Customer service and pretransaction, transaction,
and posttransaction elements

Pretransaction Elements

For example:
- Written customer service policy
  (Is it communicated internally and externally? Is it understood? Is it specific
  and quantified where possible?)
- Accessibility
  (Are we easy to contact/do business with? Is there a single point of contact?)
- Organization structure
  (Is there a customer service management structure in place? What level of
  control do they have over their service process?)
- System flexibility
  (Can we adapt our service delivery systems to meet particular customer needs?)

Transaction Elements

For example:
- Order cycle time
  (What is the elapsed time from order to delivery? What is the reliability/
  variation?)
- Inventory availability
  (What percentage of demand for each item can be met from stock?)
- Order fill rate
  (What proportion of orders are completely filled within the stated lead time?)
- Order status information
  (How long does it take us to respond to a query with the required
  information? Do we inform the customer of problems or do they contact us?)

Posttransaction Elements

For example:
- Availability of spares
  (What are the in-stock levels of service parts?)
- Call-out time
  (How long does it take for the engineer to arrive and what is the "first call"
  fix rate?)
- Product tracing/warranty
  (Can we identify the location of individual products once purchased? Can
  we maintain/extend the warranty to customers' expected levels?)
- Customer complaints, claims, etc.
  (How promptly do we deal with complaints and returns? Do we measure
  customer satisfaction with our response?)

Source: Christopher, M. (2005) *Logistics and supply chain management: Creating value-added networks,* 3rd ed. Pearson Prentice Hall, London.

There are some customer service attributes that could cost around **20% of the product cost and generate 80%** of the impact on customer perception of service performance.

That is why some companies try to define strategic positioning around service policies and develop a market-driven logistics strategy **to achieve "service excellence"** by implementing the **perfect order** concept, or **"on time** (Martinez et al., 2012)—**in full and error free"** (Christopher, 2005).

Performance Objectives: Operations Costs

Supply and Logistic Chain Management Impact on the Business's Return on Investment

The pressure in most organizations is to improve the productivity of capital. The return on investment (ROI) is the ratio between the net profit and the capital, and it can be used to measure the impact of supply and logistic chain management (Figure 3.8).

The impact of supply and logistic chain management (Figure 3.9) could be as follows:

Sales growth: By improving quality service, lost sales can be avoided and customer loyalty can be increased.

Production and logistic efficiency: Units produced per hour or the orders delivered per hour can be increased by eliminating wastes related to materials flow inbound the facility.

Stock reduction: By reducing the stock level through stock management, supplier integration, and having appropriate control over the catalog of references, working capital can be reduced.

Assets utilization: By designing flexible facilities to adjust capacity to demand or by trying to level demand, it is possible to reduce costs, thereby avoiding infrautilization.

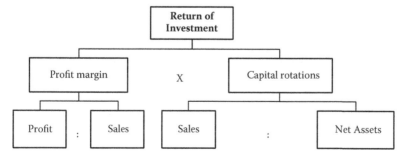

Figure 3.8 Enterprise return on investment (ROI).

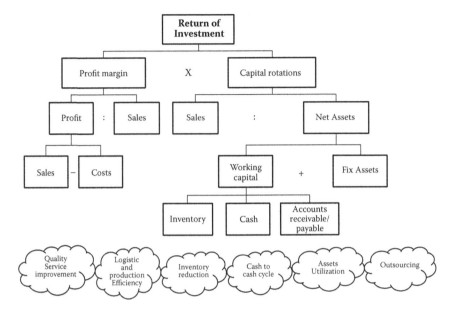

Figure 3.9 Impact of logistic management on the ROI.

Outsourcing: By outsourcing totally or partly, it is possible to guarantee operations with fewer fixed assets.

Cash-to-cash cycle: By balancing customer payments and payments to suppliers, it is possible to reduce the cash-to-cash cycle.

Total or Integral Cost Concept

Conventional accounting systems do not usually assist in the identification of cost drivers, which are absorbed in other cost elements. To evaluate logistic and supply chain management and related decisions, it is necessary to assess the cost in a systemic way by taking all the materials flow processes and, if necessary, segmenting them by customer type, market segment, or distribution channel (Christopher, 2005).

Thus, some authors (e.g., Christopher, 2005) propose a new key performance indicator called the **Logistic Total Cost.** This parameter allows the impact of decisions on different echelons in the supply chain to be evaluated (Figure 3.10 and Figure 3.11).

Cost Accounting and the Process of Internationalization

The product-market strategies represent the basis for planning and for the development of industrial production. The strategic planning process

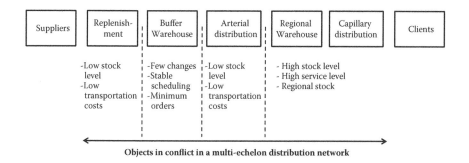

Objects in conflict in a multi-echelon distribution network

Figure 3.10 Example of dilemmas and conflicting decisions in a multiechelon supply network.

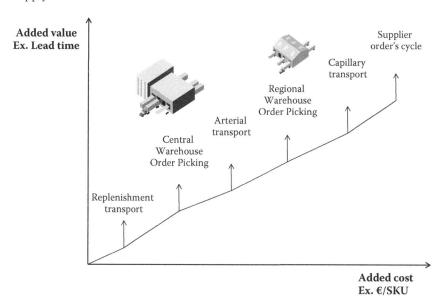

Figure 3.11 Cost added in euros/SKU and value added in time in a multiechelon supply network.

determines the key financial parameters for product development and defines the capacities and the capital investments. Strategic and performance planning *requires the development of a technical production concept,* which includes the expected product and production technologies.

The tools used to assist the **capital investment and factory performance planning should be able to** *support scenarios and evaluation* and sales planning, as well as capacity and production costs planning. These activities should be performed systematically and realistically, while taking into account the effects of rationalization and conceptual measures.

Christopher (2005) states that conventional cost accounting systems related to operations management have some difficulties when supporting decision making. Some of the difficulties worth highlighting include the following:

- Ignorance of true costs of servicing types/channels/market segments.
- Real costs information is too generic and not sufficiently detailed.
- Conventional accounting systems are functional and not oriented to output and processes.

Other authors (e.g., in the annual report of Fraunhofer, 2010) state that the advantages of these conventional systems are:

- Pinpointing actual **product costs** (for standard high-volume products and low-volume, order-specific variants) and transparency for calculating prices
- **Assignment** of **overheads** to the originating units
- **Support** in making **decisions** on high-mix/small-batch orders, technology scenarios, process optimization, and customer/market evaluations
- **Visualization** of rationalization potential in direct and indirect activities and derivation of **measures to cut costs**

Cost accounting is a tool for managers that allows them to plan and control costs by identifying cost drivers and supporting decision making. In the internationalization process, it is a strategic tool that can be useful for the subsequent decision making.

This cost accounting approach will be illustrated in two common cases: a production relocation project and a purchasing decision.

Case A: A Production Relocation Project

It would allow a rough-cut evaluation that calculated the operating costs reduction, assuming that productions will be completely relocated in one step and the additional initial investment and one-off expenses also are calculated (Figure 3.12).

Case B: A Purchasing Decision. Component Manufactured in a Dedicated Assembly Line, Decision to Move to a Low-Cost Country

This analysis could be appropriate when comparing local suppliers in a high-cost country with an offshore low-cost country's new suppliers in a global sourcing strategy or comparing new suppliers next to a new offshore facility with actual suppliers of the manufacturing network.

New Operations Cost Savings or Losses

	Material Costs	Production Costs	Supply Transport and Logistic Costs	Distribution Transport and Logistic Costs	Overhead Costs	Other Fixed Costs
Location Factors	Market Competences and Regulations Productivity	Labor Costs	Transportation Cost Rates Duty Costs Stock Increase		Salaries Local/Expatriate Blue and White Collars	
Company Sector Factors	Learning Curves	Process Technology Exploitation Costs Productivity	Agents and Means of Transport Value per Product Weigh Volume		Meeting with Suppliers/Customers/Authorities	Marketing Costs

Investments and One Off Expenses

	Investments	Startup Costs	Restructuring Costs and Physical Removal	Working Capital Costs
Location Factors	Building Machinery and Warehousing Technology	Training Costs	Ramp-up Process Cost	Supplier Payments Customer Payments
Company Sector Factors	Machinery Warehousing	Qualification and Skills	Machinery to Be Transferred Compensation (per Employee)	

Figure 3.12 Diagram of a dynamic investment analysis. (Adapted from Abele, E., et al.)

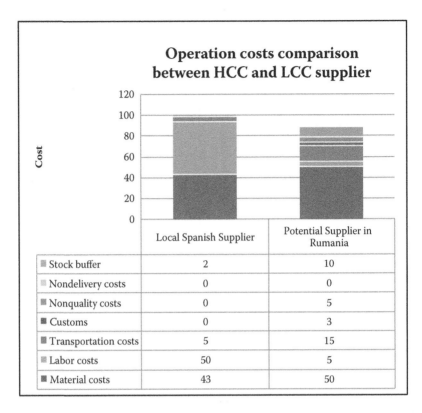

Operation costs comparison between HCC and LCC supplier		
	Local Spanish Supplier	Potential Supplier in Rumania
▨ Stock buffer	2	10
▨ Nondelivery costs	0	0
▨ Nonquality costs	0	5
▨ Customs	0	3
▨ Transportation costs	5	15
▨ Labor costs	50	5
▨ Material costs	43	50

Figure 3.13 Comparing operations costs from local medium-cost country versus new offshore low-cost country supplier.

As we can see in Figure 3.13, labor cost is an important factor, but to correctly estimate the potential benefit, other operations costs should be considered.

Nevertheless, there are some authors who are not in tune with the validity of cost accounting. The reasons are that the accounting systems do not take into account the value generated by the joining of activities (Porter, 1980), and these systems do not take into account the utilization of bottlenecks and constraints (Goldratt, 1980, 1988).

Operations Strategy and Operations Design and Management

With respect to operations resources and the related processes (see Figure 3.7), Wheelwright and Hayes (1984) state that the facility layout depends on the product volume and variety and the characteristics of the production process (Figure 3.14). Cuatrecasas (2009) pointed out that, in

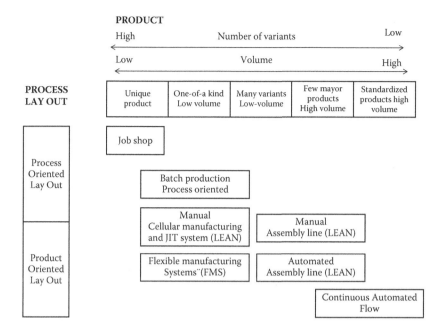

Figure 3.14 Facility layout alternatives depending on the product and process characteristics.

addition to the classical job shop, process shop, and product shop, cellular manufacturing and flexible manufacturing systems also have to be considered (see Chapter 6).

Operations Strategy and Supply Strategy

Turning to operations resources and operations capabilities (see Figure 3.7) and the associated strategy for responding to demand, two concepts can be distinguished: **Order Decoupling Point** and **Order Penetration Point**.

The lead time gap between the production lead time, i.e., how long it takes to plan, source, manufacture, and deliver a product (P), and the delivery time, i.e., how long customers are willing to wait for the order to be completed (D), is a key element of the supply chain (Simchi-Levi, Kaminsky, and Simchi Levi, 2000).

In comparing P and D, a company has several basic strategic order fulfillment options (Figure 3.15):

- **Engineer-to-Order (ETO)—(D>>P):** Here, the product is designed and built to customer specifications; this approach is most common for large construction projects and one-off products, such as ships and facilities.

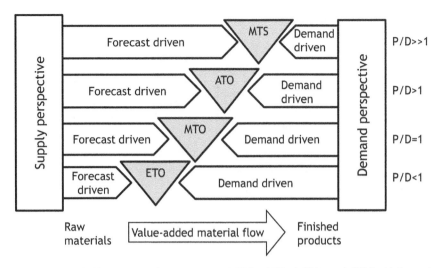

Figure 3.15 Different supply strategies: MTO, ATO, MTS, and ETO. (Adapted from Wikner, J., and Rudberg, M. (2005) Integrating production and engineering perspectives on the customer order decoupling point. *International Journal of Operations and Production Management* 25 (7): 623–641.)

- **Build-to-Order (BTO), also known as Make-to-Order (MTO)— (D>P):** Here, the product is based on a standard design, but component production and manufacture of the final product is linked to the final customer's specifications; this strategy is typical for high-end motor vehicles and aircraft.
- **Assemble-to-Order (ATO)—(D<P):** Here, the product is built to customer specifications from a stock of existing components. This assumes a modular product architecture that allows for the final product to be configured in this way; a typical example is Dell's approach to customizing its computers.
- **Make-to-Stock (MTS), also known as Build-to-Forecast (BTF)— (D = 0):** Here, the product is built against a sales forecast and sold to the customer from its finished goods stock; this approach is common in the retail sector.
- **Digital Copy (DC)—(D = 0, P = 0):** Where products are digital assets and inventory is maintained with a single digital master. Copies are created on demand, downloaded, and saved on customers' storage devices.

The Order Penetration Point (OPP) is defined as the point in the manufacturing value chain for a product where the product is linked to a specific customer order. Sometimes the OPP is called the Customer Order Decoupling Point (CODP) to highlight the involvement of a customer order. Nevertheless, it is not the same because in a fragmented

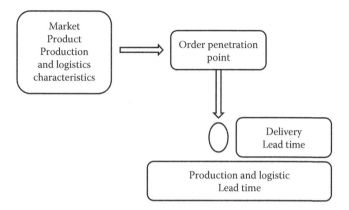

Figure 3.16 Order penetration point decision conceptual model. (Adapted from Olhager, J. (2003) Strategic positioning of the order penetration. *International Journal of Production Economics* 85 (3): 319–329.)

international materials flow there could be various CODPs, but the OPP is unique.

The positioning of the OPP is becoming a topic of strategic interest. With global markets, increasing global competition and shorter product life cycles, the choices and shifts between make-to-order (MTO) and make-to-stock (MTS) policies must be made faster and at a strategic level.

The OPP divides the manufacturing stages that are forecast-driven (upstream from the OPP) from those that are customer-order-driven (the OPP and downstream). Olhager (2003) states that the most important factors for establishing the OPP can be divided into three categories related to (1) market, (2) product, and (3) production characteristics.

Market-Related Factors

Delivery lead time, which is set by the service policy, determines how far back the OPP can be positioned.

Product demand volatility indicates to what extent it is possible or reasonable to make products to order or to stock. Low volatility means that the item can be forecast-driven.

Product volume is related to demand volumes and to what extent it is more economical to produce to stock or to order, considering integral or total cost.

A broad **product range** and a wide set of **product customization requirements** are impossible to provide on an MTS basis. The investment in final goods inventory would be too immense. However, a narrow range of products and predetermined customer choices make it possible to move toward ATO or even MTS.

Customer order size and frequency are indicators of volume and the repetitive nature of demand.

For products with highly **seasonal demand**, it may be uneconomical for the manufacturing firm to respond to all demand when it occurs. Consequently, the firm may choose to manufacture some products to stock in periods with low demand in anticipation of peak demand. Therefore, production is leveled and plant utilization increases, and a product may shift between MTS and MTO or ATO, depending upon the season.

Product-Related Factors

Modular product design is typically related to ATO product delivery strategies. Such initiatives are often a response by the producer to create a variety of choices for the customer, with a relatively short delivery lead time and manufacturing efficiency for upstream operations.

The width and depth of the **product structure** (especially in type A, see Chapter 6) indicates product complexity. A deep product structure may well correspond to long cumulative production lead times. Then, the various paths of the product structure need to be analyzed in terms of lead times to determine where in-process inventories need to be kept relative to the delivery lead time requirements.

Production-Related Factors

Production lead time is a major factor to consider with respect to the delivery lead time requirements set by the market.

The number of **planning points** in a manufacturing process restricts the number of potential OPP positions. A planning point is a manufacturing resource or a set of manufacturing resources that can be regarded as one entity from a production- and capacity-planning point of view.

The **flexibility** of the production process, e.g., through short setup times, is a prerequisite for producing to order. Therefore, a wider range of products and customization can be accommodated in the production system.

The position of the **bottleneck** of the production process relative to that of the OPP is interesting. From a resource optimization point of view, it is advantageous to have the bottleneck upstream from the OPP, so the bottleneck does not impact a volatile demand and a variety of different products. With respect to the just-in-time principle of eliminating waste, it would be best to have the bottleneck downstream from the OPP so the bottleneck only needs to be worked on on products for which the firm has customer orders. A bottleneck can be a candidate for the OPP, especially if it is an expensive resource that performs significant activities in the production process of the product (Olhager and Östlund, 1990). **Resources with sequence-dependent setup times** are best positioned upstream from the OPP. Such resources can easily turn into bottlenecks without proper sequencing, which is a likely course of action for downstream operations.

Operations Strategy: Manufacturing and Productive Sustainable Growth Models

Finally, with regard to operations resources and operations strategy decisions (see Figure 3.7), we hold that the economical and social sustainability of operations depends mostly on the adopted productive models and plant specialization.

Productive Models

Some authors (e.g., Boyer and Freyssenet, 2002) state that there are **five productive models**, each of which has different layouts, wage conditions, product strategies (variety, type of products, and volumes), **and critical success factors** (cost, variety, innovation, and adaptability to demand) (Table 3.2 to Table 3.4):

Table 3.2 Volume and volume-variety productive models

Volume:	Volume and variety:
Identify a great market niche and a production volume to reduce costs per unit.	Identify a great market segment which could be accomplished with product variety.
Product strategy:	**Product strategy:**
Identify a great market niche and a production volume to reduce costs per unit. Standard model to an affordable prize for the bigger market segment.	It is possible to obtain economy of scale through product modularity and commonality. High and low value segments are excluded. Copy the innovation of competitors.
Product organization:	**Product organization:**
Product layout with high investment in machinery and high production cadence.	Product process layout. Mass customization concept. Polyvalent and flexible machines in the areas that customized the product.
Wages:	**Wages:**
Wages higher than the region mean. Wages indexed to performance.	Wages indexed to performance and skills.

Table 3.3 Continuous cost reduction and innovation productive models

Continuous cost:	Innovation and variety:
Identify a constant market volume and exploit in a sustainable way through continuous cost reduction.	Create and translate to market innovative products that are difficult to copy or replicate in the short term.
Product strategy:	**Product strategy:**
Well equipped products for each segment. Innovations are copied once the market has accepted them.	Create new concepts and anticipate potential customers needs.
Production organization:	**Production organization:**
Lean production	The layout and equipments have to be adaptive to changes. Process flexible and low automation level, reducing the break-even point. Product process layout.
Wages:	**Wages:**
Wages indexed to production efficiency.	Wages indexed to skills and innovations in new product/processes.

Table 3.4 Variety and flexibility productive models

Variety and flexibility:

Product variety is a way of gaining market share that is served due to the system responsiveness in time and volume.

Product strategy:

Great variety of products for lots of segments

Production organization:

Process layout and flexible equipments to produce in medium batch sizes

Wages:

Wages index to responsiveness

- Volume
- Volume and variety
- Continuous cost reduction
- Innovation and flexibility
- Variety and flexibility

Plant Specialization

Another key issue when designing or redesigning the strategic capacity of a manufacturing network is the learning curves. Once the facilities have passed through the ramp-up process, they increase productivity, improve production processes, and reduce costs in a more predictable way. Sometimes the learning curves in a facility allow the competitive position of a factory to be reinforced if the production volume is guaranteed.

The focus factory concept tries to accumulate the means and resources needed to increase the effectiveness of the learning curves. When speaking of just production machinery integration, it is a volume productive model, but if other value chain activities (product engineering, process engineering, purchasing, etc.) are integrated, other productive models could be used (volume and diversity or innovation and flexibility), and the plant's role could be that of a lead factory (see Chapter 5).

The clone factory concept tries to have complementary facilities to produce the same products, and reasons for that could be:

- To exploit the cost advantage due to the facility's location in a better economic currency zone (euro, dollar, yen, etc.)
- To optimize logistic costs for each geographical area
- To exploit the cost advantage due to different cost structures or off-shore factories (see Chapter 5)

The **market factory** allows the production process in domestic markets to be customized and route-to-market costs to be optimized. The role of the plant, depending on the integrated value chain activities, could be that of a server or contributor factory (see Chapter 5).

Manufacturing Servitization

A growing number of products, in fact, are becoming commodities, while customers seek solutions rather than simple products. Durable goods industries are experiencing a trend toward the same integration of product and service that manufacturing companies offer, defined as product-centric servitization (Vandermerwe and Rada, 1988; Baines et al., 2009a; Baines et al., 2009b). Thus, some manufacturing companies are shifting their value proposition from the "sale of product" to the "sale of use" (Baines et al., 2007) in order to generate profits, growth, and increased market share.

Baines and Lightfoot (2013) adds that motives to adopt this approach include those found in Table 3.5.

Some authors (e.g., Paiola et al., 2010) state that the two key questions are the scope of servitization and capabilities acquisition (see Table 3.6) and the condition of these variables on the approach, strategic goals, and capabilities to be deployed (see Table 3.3).

Table 3.5 Motives for adopting servitization strategies

Manufacturers Motives	Customer Motives
Regulatory: To comply with legal obligations or corporate obligations. Take advantage of taxation laws and conventions	
An external focus on:	An internal focus on:
• Helping customers to improve their experience and get greater value out of equipment.	• Reducing capital investment in people and equipment and consequential fixed overhead.
• Defending revenue streams by locking out competitors who might offer services at lower cost.	• Improving financial control, smooth cash flow, and better balance outgoings with own revenue generating activities.
• Intensifying the relationship with customers and selling other products and services.	• Establishing a management team to focus energies on core business activities.
• Growing business though opening up new revenue streams with existing customers or attracting new types of customers.	• Reducing risks of acquiring and operating new technologies in the latest and most advanced equipment.
• Improving cash flow and resilience of cash flow by increasing the spread and form of revenue streams.	

Table 3.6 Four different approaches when deploying a servitization strategy

		Servitization Scope	
		Product-Based	Solution-Based
Capabilities	Internal	Product-based provision	Solution provision
Acquisition	External	Product-based services integration	Solution integration

The Scope of Servitization

The scope can be narrow when services added are strictly related to the product and aimed at complementing it in its basic and secondary functions, such as warranties, maintenance, repair, etc. On the other hand, it may (also) include services that support the customer's business processes. In this case, a broader approach to servitization is in place. This is frequently aimed at offering solutions to the customer, and it might represent the first step toward entry in a new business.

Capability Acquisition

Nordin (2005) suggests that the more strategic and customized the service offer is, the more important it is "to retain service processes internally or to align with external partners in close relationships." However, setting up a totally owned network to meet customer requirements can be very expensive, especially in cases of very large customer bases. Here again, the decision to outsource service delivery to local third parties seems to be reasonable.

Thus, given the need to integrate an extended set of competencies when developing "integrated solutions," firms do not only need to focus on relationships with end customers, but also with their business network (Table 3.7).

The typology of product service systems that companies should offer is based on two main drivers: product complexity and criticality.

The factors identified by Rapaccini, Visintin, and Saccani (2010) are:

- The service volumes and the required level of (in-house) control.
- The existing sales distribution channel(s) and product substitutability.
- The product (physical) features.
- The wish to earn direct revenue through product services.
- The cost of creating direct distribution channels, and the required degree of control over customer support quality.
- Supply chain relationships also are critical to retaining the value coming from the customer interactions, and to achieve differentiation.

Table 3.7 Four manufacturing servitization approaches, strategic goals, and capabilities to be deployed

Approach	Strategic Goal	Capabilities
Product-based service provision	Keep valuable customers and sustain core manufacturing business	• Customer relationship management training • Organizational units responsible for service not separated from the product
Solution provision	Create new complementary Business Unit	• Develop and deploy a business plan through capturing the relevant needs of customers and offering an internal developed service
Product-based service integration	Create turn-key product–service combination	• Build networks of complementary firms keeping core capabilities and linking with suppliers capabilities
Solution integration	Change the entire product–service concept	• Build networks of complementary firms and service alliances

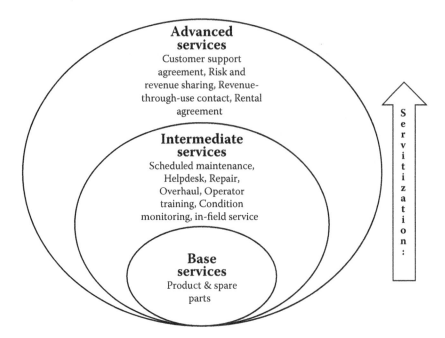

Figure 3.17 Services radar model for clustering services offerings. Baines and Lightfoot (2013).

Tukker and Tischner (2006) identifies positioning along a "product-service continuum" that ranges from traditional manufacturers merely offering services as add-ons to their products, through to service provision where services are the main part of their value creation process (Figure 3.17).

Base services are at the core of any offering from a manufacturing enterprise. These are concerned with the initial provision of equipment and associated spare parts. If an enterprise then extends into intermediate services, such as repair and overhaul, there is implicitly a greater involvement in assuring the state and condition of equipment. Intermediate services can be thought of as subsuming base services. Advanced services occur where the extended enterprise takes responsibility for the outcomes of their equipment use rather than simply the condition.

Nevertheless, the benefits and risks of advanced services are greater, and are not always profitable (Figure 3.18).

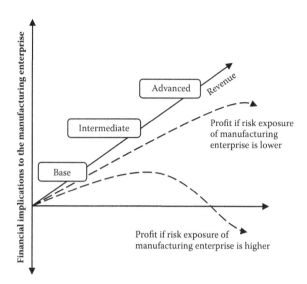

Figure 3.18 Agreed relationships between contracted risk, revenue, and profit.

GROWTH OF SERVICES—CATERPILLAR

A well-known leader of service provision is Caterpillar, the world´s largest maker of construction and mining equipment, diesel and natural gas engines, and industrial gas turbines (Baines, 2010).

The extended Caterpillar network in the U.S. demonstrates a clear and well structured portfolio of services that can be easily grouped as either base, intermediate, or advanced.

- Base services directly exploit production competences, embracing the basic equipment, such as an excavator or quarry truck, together with spare parts and consumables, plus technical support and advice.
- Intermediate services are still closely associated with production competences but their delivery requires extra resources, logistics, or organizational capacity: Field servicing, maintenance, repair, and overhaul would be typical, as would training in the correct maintenance and operation of equipment. These services can be more complex and riskier as they call for Caterpillar dealerships.
- Advanced services stretch this risk and responsibility further still and are closely associated with taking on activities that would otherwise be internal to the customer, and are reflected

in performance measures key to a customer's own business objectives. Such services are often closer to outsourcing practices than equipment supply, leading to contracts based on guaranteed levels of performance or availability. In such cases, penalties are incurred if equipment performance fails to meet very specific expectations, which in the case of Caterpillar might, for example, be the anticipated revenue from a quantity of mined ore carried by a fleet of quarry trucks.

- Fourth level of service which might include general consulting, dealing with such topics as business analytics and optimization, financial management, etc.

Caterpillar has clearly decided that this is a step too far as it could place it in competition with some of their existing customer base. To Caterpillar, the preservation of strong relationships between themselves, their dealers, and customers is critically important (Baines et al., 2011).

ENGINE HEALTH MANAGEMENT (EHM) IN ACTION AT ROLLS-ROYCE

Increasingly airlines—and now defence forces, too—want to improve their own service and minimize unnecessary costs by concentrating on what they do best and avoiding tasks they can safely leave to others (Waters, 2009). Now, around half of the Rolls-Royce civil engines fleet is covered by long-term service agreements, with 80 percent of new business incorporating long-term support elements. This trend sees airlines effectively turning over engineering and maintenance support to the engine maker and effectively buying service support "by the hour." It is important to recognize that Engine health management (EHM) is primarily there to reduce maintenance cost and avoid service disruption. This support is in addition to the well-established safety systems confirmed by engine certification and managed by the airlines.

The EHM system for Rolls-Royce aero engines comprises five key stages:

- Sense—measuring various parameters within the engine.
- Acquire—capturing this data at relevant periods during every flight.
- Transfer—transmitting data from the aircraft to the ground

- Analyse—normalizing data and detection of any unusual characteristics
- Act—providing advice to the maintainer so that corrective action can be made if necessary.

To illustrate how this works in practice, consider an incident where an engine ingests a foreign object that causes some damage to an HP compressor blade. This is unlikely to have any immediate effect on the engine operation, but the damage will steadily propagate over a number of flights until damage that is more substantial is done which will cause the engine to surge.

The EHM system monitors a number of independent parameters that can detect this before the compressor is degraded and a surge occurs. Damage to the compressor blades will reduce the efficiency, and therefore the engine will run hotter when trying to produce the same thrust. The trend of turbine gas temperature (TGT) flight by flight will therefore show a step change. Additional information from the P/T25 and P/T30 sensors (measuring pressure and temperature in front of and behind the high pressure compressor) will enable a simple calculation of the compressor efficiency to be

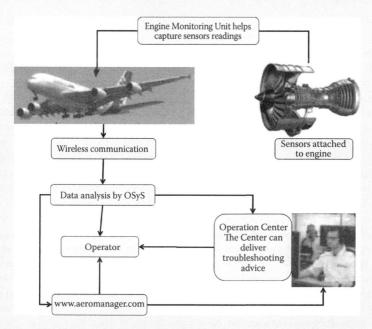

Figure 3.19 EHM system flow diagram.

made, and this is more sensitive to small changes than TGT. Even if the change in performance is too small to detect, loss of any material from some compressor blades will change the balance of the HP rotor. As this rotates at around 12,000 rpm, the vibration characteristic of the engine will change. This can be detected by the simple tracked orders captured by the ACMS, or through more sophisticated vibration signal analysis within the EMU. With multiple, independent signatures to view, the OSyS analysis system can detect that a significant change in the engine behavior has occurred. From these signatures the Rolls-Royce engineers can determine that the most likely cause is damage to the HP compressor, and request the operator to carry out an inspection by using borescope equipment or even to deploy specialized on-wing maintenance procedures—see box titled On-Wing Care.

In 2008 alone, the use of EHM mitigated 75 percent of potential civil engine in-flight events, often employing on-wing care techniques to do so.

References

Abele, E., Meyer, T., Näher, U., Strübe, G., and Sykes, R. (2008) *Global production: A handbook for strategy and implementation.* Heidelberg, Germany: Springer.

Baines, T. (2010) Growth of services. *Ingenia* 44.

Baines, T., and Lightfoot, H. (2013) *Leadership in high-value services for manufacturers.* Wiley (forthcoming).

Baines, T. S., Lightfoot, H. W., Peppard, J., Johnson, M., Tiwari, A., Shebab, E., and Swink, M. (2009b) Towards an operations strategy for product-centric servitization. *International Journal of Operations and Production Management* 29 (5): 494–519.

Baines, T., and Ball P. (2011) An operations strategy formulation methodology for manufacturing organizations seeking to adopt informated product servitized solutions. Paper presented at the European Operations Management Association (EurOMA) Conference, Cambridge, U.K., July 3–6.

Baines, T,. Lightfoot, H., Evans, S., Neely, A., Greenough, R., Peppard, J., Roy, R., Shehab, E., Braganza, A., Tiwari, A., Alcock, J., Angus, J., Bastl, M., Cousens, A., Irving, P., Johnson, M., Kingston, J., Lockett, H., Martinez, V., Micheli, P., Tranfield, D., Walton, I., and Wilson, H. (2007) State-of-the-art in product service-systems. Paper presented at the Proceedings of the Institution of Mechanical Engineers (IMechE) Part B, *Journal of Engineering Manufacture*, forthcoming.

Boyer, R., and Freyssenet, M. (2002) *The productive models: The conditions of profitability.* Palgrave/Macmillan, New York.

Christopher, M. (2005) *Logistics and supply chain management: Creating value added networks*, 3rd ed. Pearson Prentice Hall, London.

Cuatrecasas, L. (2009) *Diseño avanzado de procesos y plantas de producción flexible.* Profit Editorial, Barcelona.

Fraunhofer-Gesellschaft. (2010) Annual report. Munich, Germany.

Goldratt, E. M. (1980) Optimized production timetable: Beyond MRP: Something better is finally here. Paper presented at the APICS 23rd Annual International Conference Proceedings, Falls Church, VA.

Goldratt, E. M. (1988) Computerized shop floor scheduling. *International Journal of Production Research* 26 (3): 443–455.

Kaplan, R. S., and Norton, D. P. (2001) The strategy focused organisation: How balanced scorecard companies thrive in the new business environment. *Harvard Business School Press*, Boston.

Martinez, S., Errasti, A., and Rudberg, M. (2012) Implementing Zara's *pronto muda* paradign at a value brand retailer: an empirical study. 4th Joint World Conference on Production of Operations Management, Amsterdam, The Netherlands.

Monczka, R. M., Handfield, L. C., Guinipero, L. C., Patterson, J. L., and Walters, D. (2009) Purchasing and supply chain management. *Seng Lee Press*, Singapore.

Nordin, F. (2005) Searching for the optimum product service distribution channel: Examining the actions of five industrial firms. *International Journal of Physical Distribution and Logistics Management* 35 (8): 576–594.

Olhager, J. (2003) Strategic positioning of the order penetration. *International Journal of Production Economics* 85 (3): 319–329.

Olhager, J., and Östlund, B. (1990) An integrated push–pull manufacturing strategy. *European Journal of Operational Research* 45 (2–3): 135–142.

Paiola, M., Saccani, N., Perona, M., and Gebauer, H. (2010) The servitization of manufacturing firms: Four strategic approaches. Paper presented at the European Operations Management Association (EurOMA) Conference, Porto, Portugal, June 6–9.

Porter, M. (1980) Competitive strategy: *Techniques for analyzing industries and competitors.* Free Press, New York.

Porter, M. (1985) *Competitive advantage: Creating and sustaining superior performance.* Free Press, New York.

Porter, M. (1998) *On competition.* Harvard Business Review, New York.

Rapaccini, M., Visintin, F., and Saccani, N. (2010) Linkages between servitization strategies and sourcing decisions: A preliminary study. European Operations Management Association (EurOMA) Conference, Porto, Portugal, June 6–9.

Sainz Vicuña, J. M. (2006) *El plan de marketing en la práctica.* ESIC, Madrid.

Simchi Levi, D., Kaminsky, P., and Simchi Levi, E. (2000) *Designing and managing the supply chain: Concepts, strategies, and cases.* McGraw-Hill, New York.

Slack, N., and Lewis, M. (2002) *Operations strategy*, 2nd ed. Prentice Hall, Upper Saddle River, NJ.

Supply Chain Council (2010) Scor model. Online at: http://supply-chain.org/resources/scor

Vandermerwe, S., and Rada, J. (1988) *Servitization of business: Adding value by adding services.* Elsevier, Amsterdam.

Vereecke, A., and Van Dierdonck, R. (2002) The strategic role of the plant: Testing Ferdow's model. *International Journal of Operations & Production Management* 22 (5).

Waters, N. (2009). Engine health management. *Ingenia* 39.

Wheelwright, S. C., and Hayes, R. H. (1984) *Restoring our competitive edge—Competing through manufacturing*, John Wiley & Sons, New York.

Wikner, J., and Rudberg, M. (2005) Integrating production and engineering perspectives on the customer order decoupling point. *International Journal of Operations and Production Management* 25 (7): 623–641.

Further Reading

Baines, T. S., Lightfoot, H. W., Benedettini, O., and Kay, J. M. (2009a) The servitization of manufacturing—A review of literature and reflection on future challenges. *Journal of Manufacturing Technology & Management* 20 (5): 547–567.

Baines, T., Lightfoot H., and Swink, M. (2011b) Servitization in action: Findings from a study of the extended Caterpillar enterprise. Paper presented at the European Operations Management Association (EurOMA) Conference, Cambridge, U.K., July 3–6.

Errasti, A. (2008) *SP4 Identificación, monitorización y trazabilidad de la Cadena de Suministro. Almacén del S.XXI.* Globalog-Proyecto Integrado Científico-Tecnológico singular y de carácter estratégico (PSE) Proyecto de potenciación de la competitividad del tejido empresarial español a través de la logística como factor estratégico en un entorno global, PSS-370-000-2008-32, Valencia

Olhager, J.(2002) Supply chain management: A just-in-time perspective. *Production Planning and Control* 13 (8): 681–687.

Olhager, J., Rudberg, M., and Wikner, J. (2001) Long-term capacity management: Linking the perspectives from manufacturing strategy and sales and operations planning. *International Journal of Production Economics* 69 (2): 215–225.

Tukker, A., and Tischner, U. (eds.) (2006) *New business for old Europe—Product-service development, competitiveness and sustainability*. Greenleaf Publishing Ltd, Sheffield, U.K.

chapter 4

New Production Facilities Location and Make/Buy–Local/ Global Configuration Alternatives

Migel Mari Egaña, Donatella Corti, and Ander Errasti

> *I'm an idealist. I don't know where I'm going, but I'm on my way.*

Carl Sandburg

Contents

Introduction

In this chapter, we discuss:

- Global supply chain: Location and make or buy decisions
- Global and local facilities and suppliers configurations
- Possible manufacturing strategic and facility role of a new facility

In 1995, a consortium of leading academics and industrialists initiated a study to explore what the next generation manufacturing enterprise might look like. One of their conclusions was that several entirely new business processes would need to be developed. One of these was described as "enterprise adaptation," which is the process of systematically designing and redesigning the enterprise to cope with increasing levels of change, uncertainty, and unpredictability. Manufacturing footprint strategy has become one of those crucial new processes required for overseeing continuous enterprise adaptation.

From there, the concept of **footprint strategy** (Christodoulou et al., 2007) appeared. It is a repeatable, long-term process that needs to be embedded in annual business planning. Because new roles and responsibilities are needed at enterprise, product, and regional levels and the other new measures and mechanisms have to be created to ensure companies know whether they are succeeding, it is completely necessary that the plan is aligned with the business strategy to ensure consistency. This new enterprise adaptation process needs to be in place for 10 years or more. It will take at least this long for globalization of markets to stabilize, for infrastructures to mature, and for the fundamental footprint changes to be set in place.

New Production Facility Location and Configuration Alternatives

Once the market strategy and product strategy of these markets are defined, considering the customer service strategy and distribution complexity of demand, the productive and logistics operations' possible alternatives have to be fixed.

The global productive and logistic network design copes with:

- The location of facilities (plants and warehouses)
- Integration or fragmentation of productive and logistic operations, make or buy decision of the activities
- Manufacturing strategy and facility strategic role (market focus, product focus, etc.) in the global network design (see Chapter 5)
- Supplier network design (see Chapter 8)
- Distribution network design (see Chapter 10)

This chapter will focused on answering the questions above for new production facilities. More specifically, on the new facilities location, logistic, and productive configurations and manufacturing role. Suppliers and distribution network design will be introduced later on.

Facilities Location

Barnes (2002) states that there are two main factors in the location decision of a facility: **access to new markets** (potential market and company growth) and **access to resources** (potential availability to strategic raw materials, low cost labor or skill workforce, or researchers).

Even if there are some enterprises that have to analyze different factors to accomplish this decision, other companies have to follow the strategy and the location decisions defined by their customers' global operations strategy (follow the leader).

When deciding the location of a new facility, it is quite common to consider some cost aspects, such as labor costs. Nevertheless, there are other operation costs and ramp-up process costs that are not always considered properly. Womack and Jones (2003) cited various **connectivity costs** that should be considered:

- **Overhead costs** assigned to product in high wage countries, that do not disappear and are redistributed
- **Stock in process increase** due to long in-transit transport time
- **Security stock increase** to guarantee service due to transport unreliability
- Increase of transport costs due to urgent transports to avoid stock-outs
- **Blue collar and white collar** visits to the offshore facility to ease the ramp-up process
- Cost of product obsolescence
- Cost of generating new competitors from new suppliers

These cost drivers are not easily calculated by operations and purchasing managers, for example, when deciding to relocate a production in a low-cost country. Thus, Womack and Jones (2003) propose three new risk and cost drivers to be considered:

- **Monetary risks**, which could have a sudden and unpredictable impact
- **Political risks** of the region or nation
- Connectivity risks cited upwards

These connectivity costs increase when the distance to markets, production sites, and supplier sites are farther.

To structure the location of a new facility, MacCarthy and Atthirawong (2003) identify 13 factors to be considered (Table 4.1).

Some authors (e.g., Abele et al., 2008) state that these factors have to be considered hierarchically in the location decision process (Figure 4.1).

Moreover, it is necessary to make some assumptions about production technologies, productivity levels, and quality performance across different regions. Will the most advanced technology be used everywhere or will this be varied according to the cost of labor? Should

Table 4.1 Factors to be considered in a facility location

Main Factors	Subfactors
1 Cost	Fix costs, transport costs, wages, energy costs, manufacturing costs, construction costs, research and development costs, other costs (transaction, management, etc.)
2 Workforce	Quality, availability, unemployment rate, unions, work culture, absenteeism, rotation rates, motivation
3 Infrastructures	Transport modes (air, harbor, train,); quality and reliability of transports modes; IT communications; quality and reliability of services (energy, water)
4 Supplier proximity	Suppliers quality, alternative suppliers availability, market competence, response time, and supply reliability
5 Market/customers proximity	Sales point proximity, market volume and tendency, response time, and required quality service
6 Proximity to other facilities	Proximity to the matrix and other facilities, robust transport routes among facilities
7 Proximity of competitors	Competitors location
8 Life standard	Standard of living, environmental issues, education system, health service, security and crime, weather
9 Legal and regulatory conditions	Legal system, bureaucracy, capital transfer regulations, benefits to expatriates
10 Economic factors	Taxes, country economical and financial stability, GDP growth
11 Political factors	Government stability, foreign direct investment attitude
12 Cultural factors	Norms, culture, language
13 Regional location concrete characteristics	Attitude of the local government, availability of terrain for future expansion, weather, access to customers and suppliers, access to resources (water, energy)

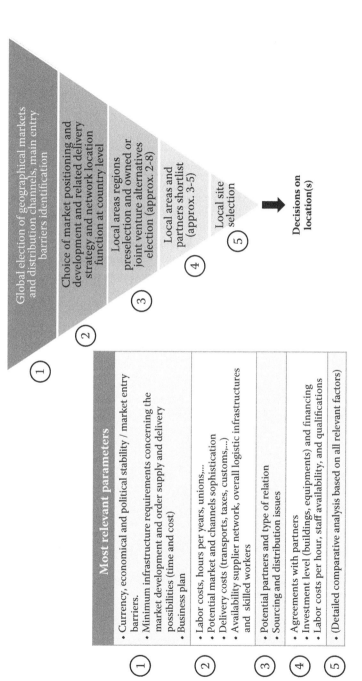

Figure 4.1 Location decision process and factors. (Adapted from Abele et al. (2008) *Global production: A handbook for strategy and implementation.* Springer, Heidelberg, Germany.)

costs be adjusted to match different productivity levels expected
in different regions or should we have the same expectation everywhere? Therefore, it is important to **state** which are the key variables
and the key assumptions. **They are listed in Table 4.2 and come from**
Christodoulou et al. (2007).

Keeping in mind the variables in Table 4.2, the world can be divided
in different geographical regions (Figure 4.2).

Table 4.2 Key variables and assumptions to locate a plant

Key Variables	Key Assumptions
Demand	Production technology
Manufacturing costs	Regional productivity
Logistics costs	Regional quality
Labor rates	Supply base maturity
Transportation time	Inventory/safety stock
Tariffs	Transition costs
Inflation	Strategic horizon
Exchange rates	

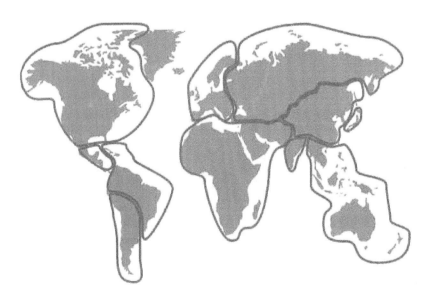

Figure 4.2 The world divided in different geographical regions. (Adapted from
Christodoulou, P., et al. (2007) *Making the right things in the right places: A struc-
ture approach to developing and exploiting "manufacturing foot print" strategy.* IFM,
University of Cambridge, U.K.)

Supply Chain Make or Buy Decision

All the activities to manufacture a product could be done in the same facility. In this case, the production has a vertical integration. Nevertheless, the concept of vertical integration has been extended to the case that the company executes all the activities even if they are in various sites in a fragmented way. In the other direction, if a manufacturer decides to produce the wide range of products to supply to the customers, it says that the integration is horizontal.

When designing a supply chain, one of the key issues is the decision to invest in the new infrastructures and equipment, and to carry the activity with one's own personnel or develop new suppliers who decide to invest and/or carry out the activity for the customer company as subcontractors. This is the so-called **make or buy decision.**

In addition to the make or buy dilemma, some authors state a combination of both, called **make some,** is better (Christodoulou et al., 2007). It could be, for example, that retaining some manufacturing processes equips us to manage the supplier interface more effectively or even that local production could have a positive marketing consequence.

A make or buy decision needs an assessment process, where different issues should be considered, such as the infrastructure potential utilization due to the business production plan, the investment capabilities, and the break-even point analysis of the activity to make the investment worthwhile.

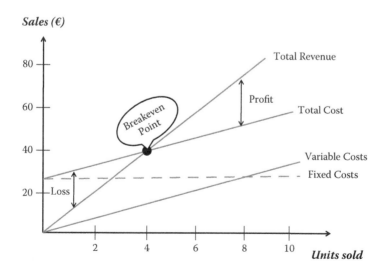

Figure 4.3 Break-even point concept.

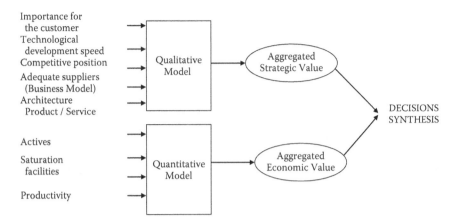

Figure 4.4 "Make or buy" decision-making tool based on aggregated economic value and aggregated strategic value of the activity. (Adapted from Fine, C. H., et al. (2002) Respuesta rápida. *Estrategia* 3 (4).)

The decision-making process has to consider **quantitative** and **qualitative aspects** and the decisions could be **to outsource totally, to outsource partially,** or to carry out the supply chain activities **with one's own means.**

Fine et al. (2002) proposed the need to balance aggregated **strategic value** and aggregated **economic value** of the potentials activities and whether or not to analyze the outsourcing (Figure 4.4).

Strategic Value

Fine et al. (2002) state that not only the economic factors are important in make or buy decisions, there also are strategic qualitative factors, such as the importance of the activity for the customer, the activity technological development speed, the existence of adequate suppliers, and the alternatives of migration in the value chain related to the sector maturity level.

Is the Activity Important for the Customer?
Is it a Core Activity or Core Business?

A core activity or core business is a joint venture of products, process, customer segments, and technologies that a company has to maintain a competitive advantage. Prahalad and Hamel (1990) state that companies should concentrate on core competences and outsource the rest of its activities to third-party logistic providers and manufacturing suppliers. If the activity is part of a core process, it should not be outsourced.

What Is the Effect of the Technological Development Speed?

If the technological development rhythm is not high, there is an entry barrier that avoids new players and competitors from entering into the market and the possibility of hypercompetition.

If the technological speed is slow in product leadership strategy, the activity should be outsourced. If the technological speed and the capital intensity are high, the company should consider making the product or developing a strategic alliance.

When the subsystem and components are standardized, the production and logistic integration is not a necessary rule and the tendency to empower the outsourcing strategy increases creating extended enterprises and suppliers networks.

As an example, Figure 4.5 presents the evolution of suppliers and the creation of module integrators (Tier 1), value-added component suppliers (Tier 2), and raw material and noncritical component suppliers (Tier 3).

When production activities are outsourced, the materials flow among the main manufacturer and manufacturing suppliers require advanced logistic and transport services. These types of services are done by third-party logistic suppliers.

Adequate Suppliers' Availability?

The existence of adequate suppliers determines the availability of obtaining competitive suppliers in terms of cost, service, and quality. Thus, it makes less sense to develop and execute the activities. When deciding the

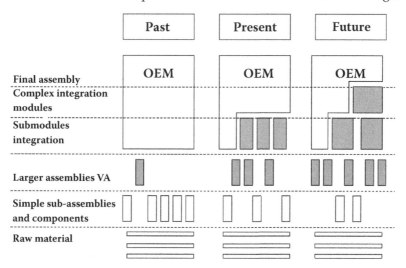

Figure 4.5 Evolution of automotive sectors supply chains.

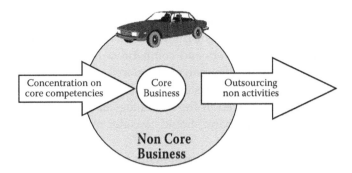

Figure 4.6 Core business concept.

suppliers, the business model of the supplier (in terms of corporate social responsibility) has to be considered.

If there are not adequate suppliers, the company could decide to develop a new supplier in order to acquire the abilities and technical skills to be a potential supplier.

Are There Profitable Alternatives of Migration in the Value Chain Related to the Sector Maturity Level?

Some companies try to change the position in the supply chain and value chain to gain a competitive advantage (Figure 4.7). This way is just an alternative because other companies that integrated vertically also have obtained the same results. This strategy is more intensive in capital.

In Figure 4.8, the decision tree and the factors of the strategic value qualitative analysis are shown.

Economic Value

Relating to economic, financial, and productivity factors, Fine et al. (2002) state that fix asset investment, equipment and personnel utilization, and internal/external productivity should be considered.

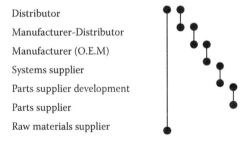

Figure 4.7 Possible supply chain migration or vertical integration opportunities to gain competitive advantage.

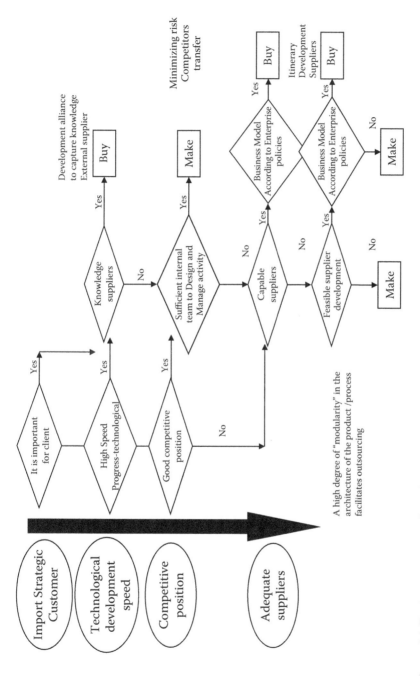

Figure 4.8 Decision tree to analyze the qualitative strategic factors.

Which Is the Fix Asset Investment Capacity?
The company that is deciding to develop a logistic activity could have a limitation in the investment capacity to accomplish new infrastructures and equipment acquisition. Besides, some assets could require a minimum return of investment.

Break-Even Point Activity Level and Risk of Demand Volatility?
Even if a new facility could be a prerequisite to enter a new market, the activity level in the area could not be enough and, to minimize risks, the demand could be volatile. In this case, the minimum investment policy is needed.

In cost strategy, the internal and external productivity also has to be considered in terms of monetary units (euros/hour) and in terms of production efficiency (units/hour).

In Figure 4.9, the decision tree and the factors of the economic value quantitative analysis are shown.

Decision

At the strategic level, the decision to develop (internal or external) is a dependent or independent decision of supply and knowledge (Prahalad and Hamel, 1990).

There are multiple options:

- Internal manufacturing
- Investment to make it internal
- Total/partial outsourcing
- Strategic alliance
- Outsource and quit internal investment
- Develop new suppliers

In Figure 4.10, the alternatives (taking into account strategic and economic value) are presented.

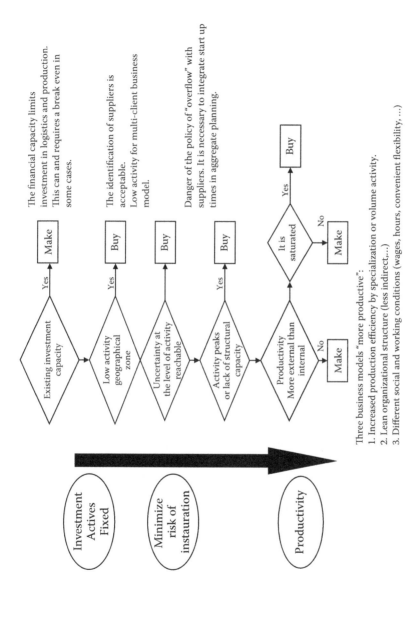

Figure 4.9 Decision tree to analyze the economic quantitative factors is illustrated.

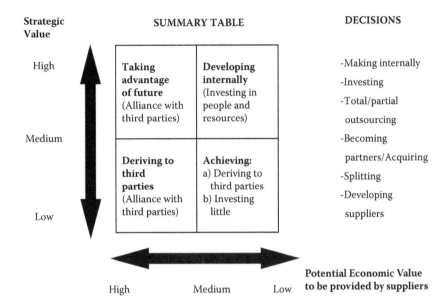

Figure 4.10 Alternatives in terms of strategic and economic value.

TUTTO PICCOLO

PRODUCTION OFFSHORING: THE OBSTACLES

Offshoring is defined as the movement of a business process done at a company in one country to the same or another company in another, different country. It is used to describe cross-border or international relocation of work, particularly in the context of relocation to lower cost areas. This relocation of work includes a company's production operations and the transfer of service activities to foreign countries in order to take advantage of a supply of skilled but relatively cheap labor.

Tutto Piccolo is a company of the Textile and Clothing Sector placed in Alcoy (Alicante), founded in 1983, although the family tradition in the textile industry dates back to 1860. The company is dedicated to the world of garments for babies and children until 8 years old. Tutto Piccolo is a familiar SME, and currently their experience comes from six generations. Almost 140 years later, Tutto Piccolo has retained all the knowledge of the trade, which has been combined with modern technology in order to be more productive.

The company has faced significant challenges over the last few years: expanding into new foreign markets and responding to increasing domestic demand. In the domestic market, products are distributed by leading multi-brand shops with a portfolio of more than 900 customers. Europe is its second market and its main destinations are Portugal, Germany, the U.K., Ireland, Austria, and Italy. But the company is not limited to Europe: Australia, Canada, the U.S., Philippines, Finland, Japan, the Middle East, South Africa, Russia, and Venezuela are just a few examples of their commitment to growth.

The offshoring strategy of Tutto Piccolo was adopted 15 years ago and, during all these years, the company has had to take decisions about where, how, when, and why to offshore its production system. The first selected country was Morocco and the offshored activities were mainly sewing. Then, the company decided to move its production to Bulgaria, offshoring the activities of cutting and sewing. However, the demands of the market and the high prices competition forced Tutto Piccolo to seek new alternatives and the company decided that the most profitable option was the Asian countries, due to the low-cost labor. Currently, Tutto Piccolo has offshored all its production processes in China, India, Bulgaria, and expanding to other countries.

Tutto Piccolo maintains its core competences and business processes, defined as those processes that are critical to achieve the success of an enterprise business, within the house-company while its production unit is offshored, hiring it abroad, in countries whose manufacturing costs are lower. Through this offshoring policy, the company can dedicate more time to what is really their business.

Table 4.3 How Tutto Piccolo faced the obstacles of the offshoring process

Obstacles	Solutions
Fear and resistance to change by human resources	Relocation of their human resources toward core business processes.
	Efficient training.
Feelings of control losses	Control structure and a proper management of the relationships with offshored suppliers.
	Development of an e-offshoring strategy.
Fear of confidentiality losses	Highly skillful labor difficult to imitate.
	Security with the information management.

(continued)

Table 4.3 How Tutto Piccolo faced the obstacles of the offshoring process
(continued)

Natural barriers (working timetables, language, culture, etc....) and geographical distance between customer and supplier	Organization and planning to be in touch the maximum hours that the partners coincide.
	Use of ICTs.
	Human resources with a great knowledge of foreign languages.
	Development of a detailed description of deadlines.
Obstacles	Solutions
Differences in the quality levels of the production process, and products and services	A complete and accurate description of quality requirements and specifications.
	Exhaustive quality control.
	Assistance and support.
Obsolescence of software systems	An area specializing in ICTs, to develop new software tools in order to provide flexibility.

Supply Chain Configuration

Supply Chain Global versus Local

The development of competitive raw materials and components markets causes the original equipment manufacturers to usually develop an outsourcing strategy (Figure 4.11).

These suppliers are classified in systems integrators (Tier 1), value-added component suppliers (Tier 2), and raw materials and noncritical component suppliers (Tier 3). The architecture of the supply chain in this case acquires a pyramid shape considering the number of suppliers. The value-added is different in each echelon (Figure 4.12).

The original suppliers' networks were usually next to the original equipment manufacturer (OEM) locations; nevertheless, the internationalization operations tendency poses multiple configurations. Meixall and Gargeya (2005) state that the raw materials, components, manufacturing, and assemblies stages could be locally or globally configured (Figure 4.13).

The following question arises: Which part of the supply chain should be globalized or localized?

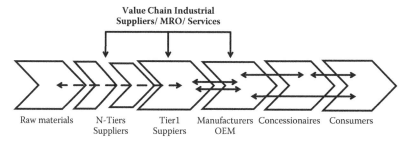

Figure 4.11 Multiagent value chains.

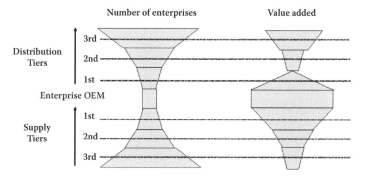

Figure 4.12 Shown are suppliers' and distributors' Tier 1, Tier 2, and Tier 3 classification, number of enterprises, and value-added components.

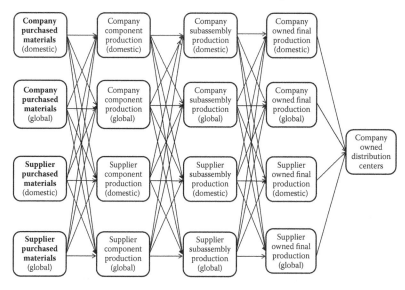

Figure 4.13 Global operations framework regarding domestic or global configuration of materials, components, subassembly, and production.

Mondragon Group's (MCC) commitment in China started a long time ago. In order to facilitate the footprint decisions in China, an industrial park was promoted and constructed with regional authorities' commitment. This provides all the Basque industry the opportunity to operate on its industrial park, in the same conditions as the companies of the Mondragon Corporation. Nowadays, MCC is extending this concept in India, exactly, in Pune and in Russia. For example, the Park Kunshan, close to Shanghai and located in a high-density industrial area, will cover a surface area of more that 500,000 m^2, providing jobs for more than 3000 people and will have involved a total investment (land, site development, building and services) in excess of 100 million Euros.

METAGRA

METAGRA is a Tier 2 supplier for automotive, aeronautic, railway sectors of a broad range of cold formed components. The company is a well recognized supplier of metallic pieces for the most important European automobile makers.

To offer an integral service for this sector requires the joint development of products with the customer, the design and manufacturing of toolings, a full manufacturing process and the delivery to the customer of a fully guaranteed product.

METAGRA has adapted its Operations configurations to cope with the new requirements of the automobile sector. For METAGRA, the most advanced degree within the automobile industry is to be considered as **Global Supplier** of very high tech components. These high tech components should be delivered on time, on full and error free wherever the customers' facilities are (e.g., Asia, America, Europe). Complementary METAGRA is also delivering an additional range of components, where high quality standards and customers'costs requirements fit together. In conclusion, METAGRA reconfigures its operations network and strategy for fulfilling the customers' requirements.

Different Supply Chain Configurations

Local Manufacturer and Supplier

The integrated production systems with local suppliers are carried out in a **synchronized production** way. The production capacities of all echelons are balanced and there is a tangible takt (pace or rhythm) time trying to optimize the materials flow in terms of quantity, response time, stock, and equipment efficiency.

These are demand-driven production systems, where replenishment system logic is implemented through pull logic and Kanban (method of inventory control developed by Toyota) techniques. If the system is completely balanced, there are no constraints, but there is a coordination point or pacemaker. If there is a bottleneck, the Drum–Buffer–Rope technique is applicable to try to exploit the bottleneck of the system, protect it with buffers, and coordinate with the rest of manufacturing system elements.

Local/Global Raw Materials and Components Suppliers

The location of suppliers does not need to be near the manufacturing plant (Figure 4.14).

The proximity of a supplier to a single manufacturing plant depends on the following factors (Figure 4.15):

- Business volume to justify the investment
- Number of supplied references or complexity
- Logistic volume total and per reference, per replenishment cycle, and production cycle
- Response time and system responsiveness

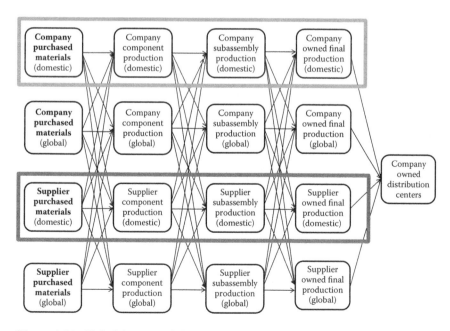

Figure 4.14 Global framework for domestic supplier and manufacturer.

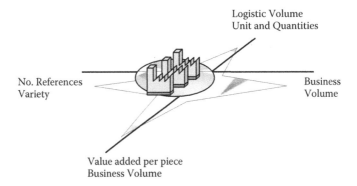

Figure 4.15 Global framework for domestic supplier and manufacturer.

Thus, the OEMs have a fragmented production system, where the suppliers' network is composed by local or domestic suppliers and offshore suppliers. These offshore suppliers need the coordination of quality control, delivery time with a long lead time gap between response time, and customer delivery time. Thus, **just-in-time (JIT) principles are not applicable in depth and a decoupling point could be needed** (Figure 4.16).

If the facilities production planning is not coordinated with the suppliers' facilities, there is a need for designing and managing a decoupling point and the corresponding stock management (see Figure 4.17).

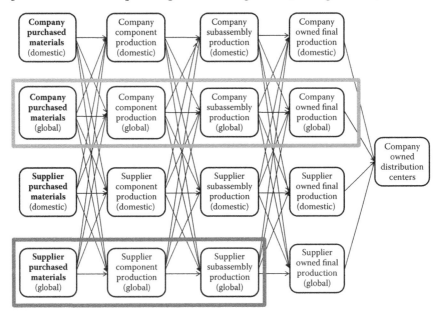

Figure 4.16 Global framework for domestic/global raw materials and purchasing.

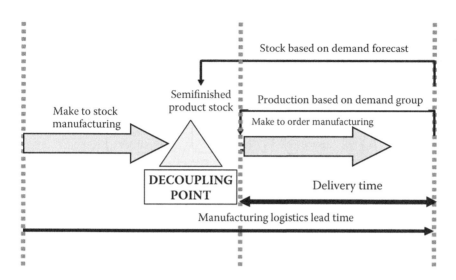

Figure 4.17 Decoupling point in nonsynchronized production systems.

Subassembly Local/Global Purchasing and Multisite Production

Some OEMs decide to fragment the internal production in various locations. It is called a **distributed production system**, which also needs coordination of the multisite network. Depending on the transit time, it is possible to develop JIT systems, e.g., engine transport from Germany to Spanish automotive assembly plants. However, if the transit time is long and the variety of references is high to cope with a high service level, there is a need to protect with stock, such as the example shown in Figure 4.16. If the logistic route is not reliable and the quality is not guaranteed, **just in case (JIC) should be applied instead of JIT.** Some companies have developed a mass customization manufacturing strategy to cope with low lead time and high variety (Figure 4.18 and Figure 4.19) (Chackelson et al., 2013).

Local/Global Supplier SMEs Operations Strategy Dilemma

There are small and medium enterprises (SMEs) whose main customers are OEMs, even if there are multinational suppliers in some sectors, such as in the automotive and aeronautic sector. OEMs have usually more resource capacity (financial, technical, etc.) to accomplish an internationalization process. SMEs are sometimes forced to start up with this process to guarantee the capabilities needed by internationalized OEMs.

Nevertheless, SMEs could have the dilemma of whether to migrate to more value-added sectors, to internationalize operations, or both.

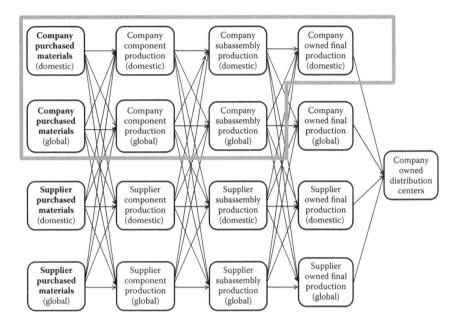

Figure 4.18 Global framework for domestic/global assembly production.

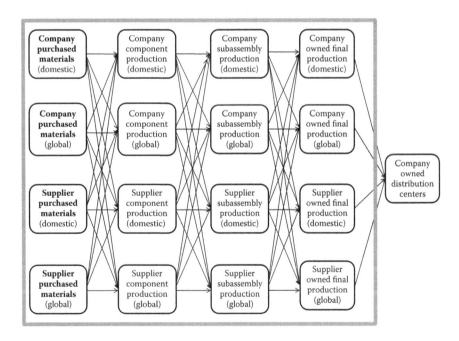

Figure 4.19 Global framework for multisite production.

Table 4.4 Generic strategies

Innovate to Generate More Value-Added Product or Services	Change the Position in the Value Chain
• New products or improvement of the existing ones	• Outsourcing. New make of buy decisions
• Manufacturing servitization or create services around products	• Integrate vertically the distribution network
• Different product and services integration	• Integrate vertically the suppliers' manufacturing network

Source: From Miltenburg, J. (2009) Setting manufacturing strategy for a company's international manufacturing network. *International Journal of Production Research* 47 (22): 6179–6203. With permission.

Migration to More Value-Added Sectors or Customers

Following the product/service leadership strategy, there are some enterprises that make a new value-added proposition, reconfiguring the value chain, or creating new processes and services for which customers are ready to pay more (Table 4.4).

International Operations Alternative Strategies

There are some operation strategies that could be carried out, such as:

1. Manufacturer–distributor strategy: It consists of manufacturing some products and commercializing other purchased ones to complete the product range offering better quality service.
2. Multisite strategy: It consists of locating facilities in different countries, with lead, offshore, and server factories (see Chapter 5) supplying domestic markets.
3. Global supplier strategy. It consists of supplying a product or service to all the needs of global markets and customers.

Manufacturing Strategy

Manufacturing strategy is how a company uses its assets and prioritizes its activities to achieve its business goals (Miller and Roth, 1994); that is to say, it is a plan for moving a company from where it is to where it wants to be. Therefore, manufacturing strategy depends on a company's industry and geographic location and is a pattern of competition that tries to generate competitive advantage (Chen, 1999).

It is possible to find different models for manufacturing strategy, but one of the last was developed by Miltenburg (2009). This model is focused on a company's international manufacturing network. The model, which is developed below, takes into account the following six objects: generic international manufacturing strategies, manufacturing networks,

IRIZAR GROUP

Irizar Group is a global project. Mainly manufactures coaches for the short, medium and long distances. Continues to strengthen as a benchmark in luxury coaches, being the market leader in Spain with a market share of 45% and an important reference worldwide.

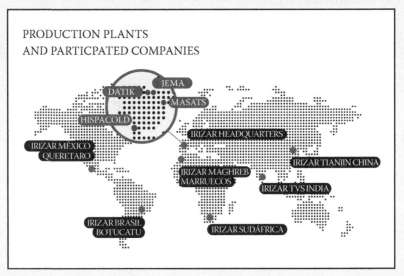

In addition to the headquarters of the company, Irizar S. Coop. in Ormaiztegi (Spain) since 1889, has production plants coaches in Brazil, Mexico, Morocco, South Africa and India (100% subsidiaries in all countries but in India it is a joint-venture) and commercial presence in more than 90 countries on five continents. Currently, 80% of sales come from outside, and over 75% of its employees working in foreign plants. Irizar Group is completed by other companies that manufacture components for industrial vehicles (International Hispacold and Masats), and more recently, as a result of industrial diversification strategy to strengthen its competitive position, is present in sectors of electronics and communications (Jema and Datik).

In all cases, the coach production plants have been carried out considering profitable growth in markets that were not accessible, with profitability since Ormaiztegi. From the technological point of view, in all cases Irizar has been implemented in countries with less developed technological sector, reinforcing the idea that market is sought, not technology.

CONFIGURATION AND SYNERGY IN THE
IRIZAR GROUP VALUE CHAIN

In the group's plants, production, engineering adaptation, marketing, customer service and people management are fully decentralized in the day to day, but the criteria and strategies designed from the headquarter in the implantation. Financial control is carried from the matrix and procurement and logistics are coordinated in terms of the opportunities in every country, always ensuring the product quality criteria defined.

The R&D is performed in the parent plant, which coordinates the launch of new products. The design and development of new coaches are done with the participation of collaborators, suppliers, customers, drivers and passengers with an analysis of their needs. All production plants, the parent and the subsidiaries have a common catalog of models. However, each country follows its own "product life cycle" according to the local market, in each country the subsidiary decides what kind of coach models offer and therefore produce. Although the process of creating new models leads from the central cooperative, also encourages foreign factories launch their own design adaptation projects, if they have a particular need based on their target market. Here too, the centers are increasingly autonomous, and all plants have R&D for the Adaptive Design.

Following internalization and market knowledge found strong synergies for the Group, mainly in procurement of materials and components. Irizar has an extensive network of suppliers, global and domestic, that are essential parts of the value chain to realize customer loyalty. Regarding major suppliers (integrated suppliers) provides what may be called a "logistics management by coach," as these suppliers are integrated so that they can consult the management system of Irizar production planning, pending orders, merchandise receptioned and continuous customer reviews regarding specifications confirmed and their associated articles by coach.

This organization also allows sharing synergies for productive capacities and the possibility of combined manufacturing between different plants of the group, thanks to the trunk of the body of the coach is the same in some of its plants, having peculiarities by country. This is extremely useful, allowing in a plant producing coaches to meet the needs of customers of another, balancing load/capacity if necessary. And while remaining very competitive delivery times to customers.

In the past, the possibility of products exchange was very high, because the product was slightly customized for each country. The present reality is not the same, but this is not deny that we continue having synergies between some plants and some products and, in the highest degree, in some components and subassemblies.

For example, when the Ormaiztegi plant was saturated with European market demands, the factory of Brazil produced CBU format (basically finished and just to personalize details in the target plant) to mount the final details of the Spain plant. Similarly, the Ormaiztegi central manufactured coaches on chassis produced from Mexico to be able to meet a peak demand of that market. Today in South Africa plant, for example, are mounted PKDs (unassembled complete bodywork to the chassis) from Brazil, because in this country the wheel goes right, as in Britain.

Managing this Extensive Value Chain (from suppliers to customers suppliers, including all the group companies), in addition to being an important source of competitive advantage has been the subject of many external recognitions:

- 2006: ICIL Award for Logistics Excellence,
- 2008: Coordination of Supply Chain and Logistics recognized as good practice by Mondragon Corporation,
- 2011: Awards for his multilocation of Good Practice in Corporate Social Responsibility, by the consortium of Emmaus Social Foundation, MIK-management Innovation Center at the University of Mondragón and GAIA Cluster:
  - Responsible management of purchasing in Morocco
  - Supplier development: practices for the integration of people at risk of social exclusion in India

Mentxu Baldazo
Global Sustainable Competitiveness Coordinator
Ex – Global Supply Chain Coordinator
www.irizar.com

IKOR GROUP has offered integral Electronic Development, Engineering and Manufacturing Services (EMS) to the industry since 1981, treating customers as technological partners from the Development of the Project through to the serial production, providing added value and competitiveness to the end product.

Headquarters are located in San Sebastian (Spain), where the lead factory is located and the IKOR TECHNOLOGY CENTRE, the aim of which is to provide technological support to each customer. The R&D&I Unit, organized in technological divisions, is complemented with support areas such as Technical Office, Product Engineering, Technological Surveillance, together with the Customer Team manage the technical-commercial relationship with customers. A complete Laboratory, equipped with updated testing equipments, allows validating and pre-certifying new developments.

Attending the demands of the global market, IKOR GROUP has set up production plants in Europe (San Sebastian–Spain), America (Guadalajara–Mexico) and Asia (Suzhou–China), in order to manage worldwide the whole Supply Chain.

At each of its plants, IKOR GROUP provides customers with the latest state-of-art production means and know-how, following the fast evolution of the electronic market, and based on an efficient investment policy.

The basis of firm and constant growth over time is based on the strategy in market diversification (Automotive, Elevators, Home Appliances, Medical, Energy, Transportation) and a Quality strategy based on a Process Management System to guarantee the ongoing improvement of all its activities.

network manufacturing outputs, network levers, network capability, and factory types.

Generic International Manufacturing Strategies

Strategies for international manufacturing are often viewed as responses to two competitive pressure that companies face. On the one hand, there is the pressure for globalization that is the necessity that exists for a company to design, manufacture, and market products on a worldwide basis. This pressure comes from five forces (Porter, 1985; Thompson and Strickland, 2001) within a company's industry: actions of competitors, bargaining power of customers, bargaining power of suppliers, threat of new competitors, and threat of substitute products.

On the other hand, there is the pressure for local responsiveness that is the necessity that exists for a company to adapt its practices to meet the varying requirements of customers, employees, and governments.

Therefore, the relation between these two pressures gives as a result seven generic strategies as shown in Figure 4.20.

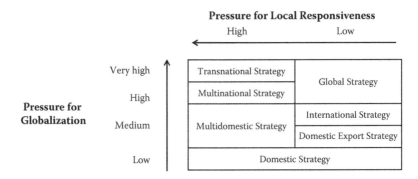

Figure 4.20 Generic strategies for international manufacturing. (From Miltenburg, J. (2009) Setting manufacturing strategy for a company's international manufacturing network. *International Journal of Production Research* 47 (22): 6179–6203. With permission.)

Manufacturing Networks

Manufacturing networks "are composed of interorganizational ties that are enduring...and include strategic alliances, joint ventures, long-term buyer–suppliers partnerships, and a host of similar ties" (Gulati et al., 2000, p. 203). There are nine well-known manufacturing networks classified as "simple networks" and "complex networks" (Miltenburg, 2009):

- Simple networks: Networks that disperse facilities to national and regional locations and are appropriate when pressure for globalization is low to medium. They are called domestic, domestic export, international, and multidomestic. For example, a domestic network manufactures in one national (i.e., domestic) area for customers in that domestic market.
- Complex networks: Networks that disperse facilities to multinational and worldwide locations and they are appropriate when pressure for globalization is high to very high. They are called multinational, global product, global functional, global mixed, and transnational.

Network Manufacturing Outputs

Depending if you focused on factory (Miltenburg, 2005) or network level (Shi and Gregory, 1998), there are different strategic outputs (Table 4.5).

Network Levers

According to Miltenburg (2009), a manufacturing network can be divided into structural and infrastructural areas (Table 4.6).

Table 4.5 Manufacturing outputs provided by a manufacturing network

Source	Manufacturing Output	Definition
Factory	Cost	Cost of material, labor, overhead, and other resources used to produce a product.
	Quality	Extent to which materials and activities conform to specifications and customer expectations, and how high or difficult the specifications and expectations are.
	Delivery time and delivery time reliability	Time between order taking and delivery to the customer. How often are orders late and, when they are late, how late are they?
	Performance	Product's features and the extent to which the features permit the product to do things that other product cannot do.
	Flexibility	Extent to which volumes of existing products can be increased or decreased to respond quickly to the needs of customers.
	Innovativeness	Ability to quickly introduce new products or make design changes to existing products.
Network	Accessibility	The ease of access a company has to present and future market segments, factors of production, and government agencies.
	Thriftiness	The ability of a company to achieve economies of scale and avoid duplication of activities.
	Mobility	The ease with which a company can transfer products, processes, and personnel between facilities to new locations, and change production volumes.
	Learning	The ability of a company to learn about cultures and needs of customers, workforces, and governments, as well as process technology, product technology, and management systems, and the ease with which this knowledge is shared.

Source: From Miltenburg, J. (2009) Setting manufacturing strategy for a company's international manufacturing network. *International Journal of Production Research* 47 (22): 6179–6203. With permission.

Table 4.6 Eight areas that comprise a manufacturing network

Structural Areas	Facility characteristics	Types of facilities in a network and their characteristics, such as size, focus, and capabilities.
	Geographic dispersion	Where value systems activities are dispersed around the world.
	Vertical integration	The extent to which a network contains facilities engaged in upstream activities involving sources of supply and downstream activities involving customers.
	Organization structure	How facilities, departments, and personnel are organized in the network.
Infrastructural Areas	Coordination mechanisms	The managerial systems and computer systems that organize data, make information available, and plan, monitor, and control activities.
	Knowledge transfer mechanisms	The systems that transfer product knowledge and process knowledge between facilities and departments in a network.
	Response mechanisms	The systems and procedures that recognize, analyze, and act on threats and opportunities arising anywhere in a network.
	Capability building mechanisms	The systems and procedures that create, sustain, and improve capabilities in areas such as design, production, and service.

Source: From Miltenburg, J. (2009) Setting manufacturing strategy for a company's international manufacturing network. *International Journal of Production Research* 47 (22): 6179–6203. With permission.

Network Capability

The overall level of capability of a network depends on the level of capability of each level in the network. The different levels from low level to high level are (Miltenburg, 2009):

- Infant level: This is the level of new manufacturing networks.
- Industry average: This level is the result of 5–10 years of experience and improvement activities.
- Adult: This is achieved by a company's determined effort to improve and become an industry leader.
- World class: A network with this level of capability is among the best in the world.

Factory Types

Factories can be categorized in several ways. On the one hand, factories can be home based (located near company headquarters in a company's home country), domestic (located in the home country away from company headquarters), or foreign (located outside the home country), depending on where they are located (Voss and Blackmon, 1996).

On the other hand, factories can also be one of six factory types: server, outpost, offshore, contributor, lead, or source (Ferdows, 1997). This second classification is considered a key decision by the authors; therefore, this topic will be explained below.

Strategic Facility Role

Among the various reasons that lead to setting up an offshore or overseas production plant, three are the main drivers commonly cited in literature and tested by means of empirical investigations: access to low-cost production input factors, access to local knowledge and technology, and proximity to market (Pongpanich, 2000). Nonetheless, at first sight, it seems that the most important factor everyone mentions as far as internationalization is concerned is the first one, the reduction of cost. This myopia that considers only partially the whole phenomenon is due to the fact that the advantages in terms of costs reduction are more easily quantifiable and measurable (Englyst et al., 2005).

In fact, a common internationalization approach, which looks for a short-term cost reduction and competitiveness, is the establishment and management of foreign plants to benefit only from tariff and trade concessions, cheap labor, capital subsidies, and reduced logistics costs. Therefore, a limited range of work, responsibilities, network participation, and resources is assigned to those factories (Ferdows, 1997).

Nonetheless, **other companies demand much more from their foreign factories** and, as a result, try to get much more out of them. This approach provides not only an access to the already mentioned cost-oriented incentives, but also a globally distributed production system with much higher proximity to potential regions, with close access to customers, suppliers, or specifically a skilled, talented, and motivated workforce. Those factories have a wider range of responsibilities beyond mere production work, e.g., product or process engineering, purchasing decisions, after-sales service, etc. (Ferdows, 1997).

Ferdows distinguishes the expected benefits in tangible ones and intangibles ones. The first category includes advantages related to the reduction of costs, being those logistics costs, production costs, or labor costs.

Intangible benefits, on the other hand, considers the possibility to acquire knowledge from research centers, customers, or suppliers; to improve the level of customer satisfaction; to reduce risks related to different currency; or to look for alternative sources of purchasing. Many authors state that companies considering expanding their network by strengthening the existing network or closing plants should consider all of these elements at the same time (Ferdows, 1997; Dunning, 2000; Pongpanich, 2000; Vereecke and Van Dierdonck, 2002). Ferdows highlights that, often, companies lack this integrated vision when they deal with the global manufacturing strategies and, thus, they lose the opportunity to fully exploit the strategic potential of an offshore plant.

Of course, not all the advantages are achieved at the same time; what is important is that the company is aware of the alternative opportunities that go beyond the costs reduction and it pays attention to the strategic evolution of the plant role over time.

An interesting proposal has been done by Ferdows (1997), who proposes a classification of the strategic role of plants based on two dimensions: the **"strategic reason for establishing and exploiting the plant"** and the **"site competence."**

As for the first dimension, there are three main reasons:

- **Access to low-cost production**: Exploitation of low-cost labor is the most important reason in this respect, followed by the proximity to cheap raw materials and cheap energy.
- **Proximity to market:** The exploitation of a plant in a foreign nation allows more rapid and more reliable product delivery, and facilitates the customization of the product according to customer requirements.
- **Use of local technological resources**: Proximity to universities, research centers, and competitors, allows a company to acquire local technological know-how and skilled employees.

The **"site competence"** dimension takes into consideration the extent to which some competencies are present in the plant production process, technical maintenance, procurement, local logistics, and product and process development.

Ferdows states that the strategic role of a plant is subjected to an evolution led by both endogenous and exogenous factors. Among the main endogenous elements worth mentioning is the increasing knowledge the company gains concerning the offshore context and the management awareness. Increasing salaries, improvement of the local markets, elimination of tariff barriers, or setting up of industrial districts are some examples of possible exogenous factors that make it necessary to adapt the strategic role of a plant.

To change the strategic role does not mean only to achieve a different set of advantages, but also the need to assign new competences and responsibilities. According to these considerations and based on the two aforementioned dimensions, Ferdows identifies six types of plants

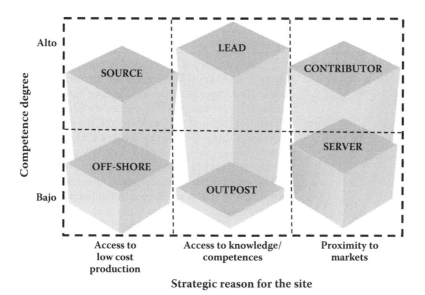

Figure 4.21 Strategic matrix and plant role classification based on strategic reason for the site and the site competence. (From Ferdows, K. (1997) *Making the most of foreign factories.* Harvard Business Review March–April: 73–88. With permission.)

(Figure 4.21) labeled as: offshore, source, server, contributor, outpost, and lead plant (see also Table 4.7).

Continuing with the manufacturing strategy model developed by Miltenburg (2009), the Ferdows model has been improved. Therefore, Miltenburg states that (Figure 4.22):

- The low roles of factory (offshore, outpost, and server) have a narrow scope of activities (factories only engage in production activities) and a low level of factory capability (these factories are commonly used in simple manufacturing networks).
- The high roles of factory (source, leader, contributor) have a broad scope of activities (factories also engage in sourcing, distribution, product and process design, and research and development activities) and a high level of factory capability (these factories are commonly used in complex networks).

Table 4.7 Characteristics of the role of plants proposed by Ferdows

Role of the Plant	Characteristics
OFFSHORE	An offshore factory is established to produce specific items at a low cost. These items are then exported either for further work or for sale. Investments in technical and managerial resources are kept at the minimum required for production. Little development or engineering occurs at the site. Local managers rarely choose key suppliers or negotiate prices. Accounting and finance staffs primarily provide data to managers in the home country. Outbound logistics are simple and beyond the control of the plant's management.
SOURCE	The primary purpose for establishing a source factory is low-cost production, but its strategic role is broader than that of an offshore factory. Its managers have greater authority over procurement (including the selection of suppliers), production planning, process changes, outbound logistics, and product-customization and redesign decisions. A source factory has the same ability to produce a product or a part as the best factory in the company's global network. Source factories tend to be located in places where production costs are relatively low, infrastructure is relatively developed, and a skilled workforce is available.
OUTPOST	An outpost factory's primary role is to collect information. Such a factory is placed in an area where advanced suppliers, competitors, research laboratories, or customers are located. Because every factory obviously must make products and have markets to serve, virtually all outpost factories have a secondary strategic role—as a server or an offshore, for example.
LEAD	A lead factory creates new processes, products, and technologies for the entire company. This type of factory takes into account local skills and technological resources not only to collect data for headquarters, but also to transform the knowledge that it gathers into useful products and processes. Its managers have a decisive voice in the choice of key suppliers and often participate in joint development work with suppliers. Many of its employees stay in direct contact with end customers, machinery suppliers, research laboratories, and other centers of knowledge; they also initiate innovations frequently.

(continued)

Table 4.7 Characteristics of the role of plants proposed by Ferdows (continued)

Role of the Plant	Characteristics
SERVER	A server factory supplies specific national or regional markets. It typically provides a way to overcome tariff barriers and to reduce taxes, logistics costs, or exposure to foreign-exchange fluctuations. Although it has relatively more autonomy than an offshore plant to make minor modifications in products and production methods to fit local conditions, its authority and competence in this area are very limited.
CONTRIBUTOR	A contributor factory also serves a specific national or regional market, but its responsibilities extend to product and process engineering as well as to the development and choice of suppliers. A contributor factory competes with the company's home plants to be the testing ground for new process technologies, computer systems, and products. It has its own development, engineering, and production capabilities. A contributor factory also has authority over procurement decisions and participates in the choice of key suppliers for the company.

Source: Adapted from Ferdows, K. (ed.) (1989) Mapping international factory networks. In *Managing international manufacturing.* Elsevier Science, Amsterdam; Ferdows, K. (1997) Making the most of foreign factories. *Harvard Business Review* March–April: 73–88.

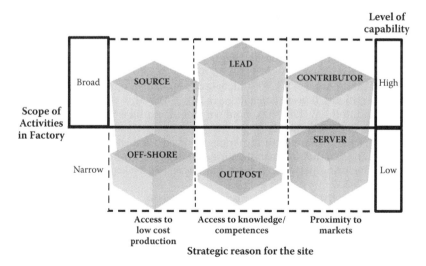

Figure 4.22 Ferdows model improved by Miltenburg. (From Miltenburg, J. (2009) Setting manufacturing strategy for a company's international manufacturing network. *International Journal of Production Research* 47 (22): 6179–6203. With permission.)

References

Abele, E., Meyer, T., Näher, U., Strube, G., and Sykes, R. (2008) Global production: A handbook for strategy and implementation. Springer, Heidelberg, Germany.

Barnes. D. (2002) The complexities of the manufacturing strategy formation process in practice. *International Journal of Operations & Production Management* 22 (10): 1090–1111.

Chackelson, G., Errasti, A., Martinez. S., and Santos, J. (2013) Supply strategy configuration in fragmented production system. *Journal of Industrial Engineering and Management* (forthcoming).

Chen, W. (1999) The manufacturing strategy and competitive priority of SMEs in Taiwan: A case survey. *Asia Pacific Journal of Management* 16: 331–349.

Christodoulou, P., Fleets, D., Hanson, P., Phaal, R., Probert, D., and Shi, Y. (2007) Making the right things in the right places. A structured approach to developing and exploiting "manufacturing footprint" strategy. IFM, University of Cambridge, U.K.

Dunning, J. H. (2000) The eclectic paradigm as an envelope for economic and business theories of MNE activity. *International Business Review* 9 (2): 163–190.

Englyst, L., Hoé Seiding, C., Wong, C. Y., and Saxtoft, M. (2005) Global production: Is it all about cost? Paper presented at the EurOMA International Conference on Operations and Global Competitiveness, Budapest, June 19–22.

Ferdows, K. (ed.) (1989) Mapping international factory networks. In *Managing international manufacturing*, pp. 3–21, Elsevier Science Publishers, Amsterdam.

Ferdows, K. (1997) Making the most of foreign factories. *Harvard Business Review* March–April: 73–88.

Fine, C. H., Vardan, R., Pethick, R., and El-Hout, J. (2002) Respesta rápida. *Gestión de Negocios* 3 (4).

Gulati, R., Nohria, N., and Zaheer, A., (2000) Strategic networks. *Strategic Management Journal* 21: 203–215.

MacCarthy, B. L., and Atthirawong, W. (2003) Factors affecting location decisions in international operations: A Delphi study. *International Journal of Operations & Production Management* 23 (7): 794–818.

Meixall, M., and Gargeya, V. (2005) Global supply chain design: A literature review and a critique. *Transportation Research Part E* 41: 531.

Miller, J., and Roth, A. (1994) Taxonomy of manufacturing strategies. *Management Science* 40 (3): 285–304.

Miltenburg, J. (2005) Manufacturing strategy, 2nd ed. Productivity Press, New York.

Miltenburg, J. (2009) Setting manufacturing strategy for a company's international manufacturing network. *International Journal of Production Research* 47 (22): 6179–6203.

Pongpanich, Ch. (2000) *Manufacturing location decisions. Choosing the right location for international manufacturing facilities.* University of Cambridge, U.K.

Porter, M. (1985) *Competitive advantage: Creating and sustaining superior performance.* The Free Press, New York, Chap. 1–4.

Prahalad, C., and Hamel, G. (1990) The core competence of the corporation. *Harvard Business Review*, May–June, 68 (3): 79–91.

Shi, Y., and Gregory, M. (1998) International manufacturing networks to develop global competitive capabilities. *Journal of Operations Management* 16: 195–214.

Thompson, A., and Strickland, A. (2001) Strategic management: Concepts and cases. McGraw-Hill Irwin, Boston, Chap. 3–10.

Vereecke, A., and Van Dierdonck, R. (2002) The strategic role of the plant: Testing Ferdows' model. *International Journal of Operations & Production Management* 22 (5).

Voss, C., and Blackmon, K. (1996) The impact of company origin on world class manufacturing: Findings from Britain and Germany. *International Journal of Operations and Production Management* 16 (11): 98–115.

Womack, J. P., and Jones, D. T. (2003) *Lean thinking.* Simon Schuster, New York.

chapter 5

Multisite Network Configuration and Improvement

Miguel Mediavilla and Torbjørn Netland

> *I can fall, I can injure myself, I can break, but it will not diminish my will.*
>
> **Mother Teresa of Calcutta**

Contents

Introduction

In this chapter, we discuss:

- The strategic role of the production plant within a network
- The Akondia framework:
  - Assessment of the strategic role of plants
  - Roadmap deployment for upgrading the strategic role
- Multisite improvement programs based on Lean management

The Strategic Role of the Production Plant within a Network

A common internationalization approach, which looks for a short-term cost reduction and competitiveness, is the establishment and management of foreign plants to benefit only from tariff and trade concessions, cheap labor, capital subsidies, and reduced logistics costs. Therefore, a limited range of work, responsibilities, network participation, and resources is assigned to those factories (Ferdows, 1997).

As stated in Shi and Gregory (1998), the multidomestic approach to the network management characterized by a weak coordination that involves the development of more or less autonomous manufacturing units geographically located close to target markets is not enough anymore to succeed in a global environment.

Evolving from an independently managed (or with lower interaction) plant network to a coordinated manufacturing network allows them to benefit from the synergy among the plants (Dubois, Toyne, and Oliff, 1993; Shi and Gregory, 2005) by improving cost and delivery performance and enhancing the learning curve from the experiences of partners in the network (Flaherty, 1986). The more advanced production networks require a global coordinated or integrated approach (Cheng, Farooq, and Johansen, 2011), where each plant has a role within the network and there are some possible paths of network optimization (Figure 5.1). This approach provides not only an access to the already mentioned cost oriented incentives, but also a globally distributed production system with much higher proximity to potential regions, with close access to customers, suppliers, or specifically a skilled, talented, and motivated workforce. These factories have a wider range of responsibilities beyond a mere production work, e.g., product or process engineering, purchasing decisions, after-sales service, etc.

What is needed is a globalized approach for the network design and management where the manufacturing system is seen as a unified whole with a mechanism of sharing knowledge, with elements of the task being performed in the most advantageous areas. An international manufacturing

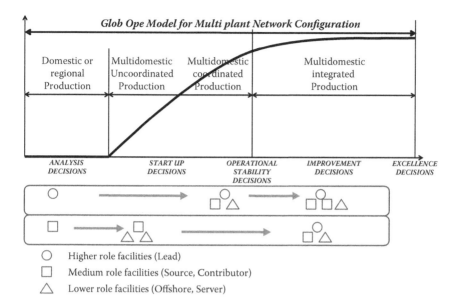

Glob Ope Model for Multi plant Network Configuration

| Domestic or regional Production | Multidomestic Uncoordinated Production | Multidomestic coordinated Production | Multidomestic integrated Production |

ANALYSIS DECISIONS   *START UP DECISIONS*   *OPERATIONAL STABILITY DECISIONS*   *IMPROVEMENT DECISIONS*   *EXCELLENCE DECISIONS*

○   Higher role facilities (Lead)
□   Medium role facilities (Source, Contributor)
△   Lower role facilities (Offshore, Server)

Figure 5.1 GlobOpe model for multisite network optimization.

system thus may be seen as a factory network with matrix connections, in contrast to the linear system of a factory.

The question that arises is **how to deploy the Operations Strategy in a multilocation Global Manufacturing Network** (GMN), i.e., how to balance the different competences and responsibilities along the different factories or facilities, taking into account that the different units of the GMN could assume different strategic responsibilities for themselves or for the whole GMN.

This approach requires analyzing, defining, and upgrading the strategic role of manufacturing and production facilities. Ferdows (1997a), who stated that the management of GMN could be executed based on the strategic plant role concept, proposes a classification of the strategic role of production plants based on two dimensions: (1) **the strategic reason for establishing and exploiting the plant** and (2) **the site competence.** According to these considerations and based on the two aforementioned dimensions, Ferdows identifies six types of plants (see Chapter 4) labeled as: offshore, source, server, contributor, outpost, and lead plant (Table 5.1).

In a recent publication, Feldmann (2011) suggests dividing the competencies into three levels instead of the two levels proposed by Ferdows. He identifies three types of plants based on the level of competence. These include: Type 1 plants that have only competence of production related matters; Type 2 plants that have competences for production and supply chain activities; and Type 3 plants that have competencies for a whole

Table 5.1 Strategic plant roles in global operation networks

		Access to low-cost production	Access to skills and knowledge	Proximity to market
Site competence	High	**Source** Low cost production Autonomy for procurement, supplier selection, production planning, outbound logistics, process changes, and product customization Same production capability as best plant in the network Location requires developed infrastructure and availability of skilled workforce	**Lead** Creates new products, processes, and technologies for the entire company Taps into local skills and technological resources to develop and transform technology into new products and processes Autonomy for supplier selections, customer communication, procurement of machinery, management of research, and innovation activities	**Contributor** Supplies national/regional market Autonomy for product/process engineering, procurement, and development of suppliers Competes to be testing ground for new process technologies and products Includes development, engineering, and production capabilities
	Low	**Offshore** Low-cost production Minimal investment in technical and managerial resources Little engineering development work Low autonomy to choose suppliers, negotiate prices, and plan logistics	**Outpost** Primary role is to collect information about suppliers, customers, competitors, and research expertise Usually has secondary strategic role as a server or offshore	**Server** Supplies national/regional markets Overcome tariff barriers, reduce taxes, logistic costs, and exposure to currency fluctuations Limited autonomy (only for decision to fit local conditions)
		Access to low-cost production	**Access to skills and knowledge**	**Proximity to market**
		Location advantage		

Source: Ferdows, K. (1997) Making the most of foreign factories. *Harvard Business Review* March-April: 73–88. With permission.

range of technical activities related to production, supply chain, and product/process development.

As illustrated in Table 5.2, several authors have proposed classifications for defining the strategic roles of plants within a GMN (Bartlett and Ghoshal, 1989; Jarrillo and Martinez, 1990; Ferdows, 1997; Veerecke et al., 2006).

While the classifications by Bartlett and Goshal (1989) and Jarrillo and Martinez (1990) provide useful insight into strategic roles of different types of subsidiaries within multinational organizations, it is Ferdows' work that specifically focuses on roles for manufacturing plants in a global network, which is the best fitting model for the purpose of this book.

Complementary to Ferdows' work, the classification of plants in global manufacturing networks done by Vereecke et al. (2006) enriches the understanding of some key competences for a global network. It is focused primarily on the intangible knowledge network and secondarily on the physical logistic networks. The authors identify four types of plants with different network roles defined as:

- **Isolated plants:** Few innovations reach the plant, few innovations are transferred to other units, few manufacturing staff people come to visit such a plant, few manufacturing people from this plant go visit other plants, and there is little communication between the manufacturing staff people of this plant and the other manufacturing managers in the network.
- **Receivers:** Like the isolated plants, but it receives more innovations from the other units in the network.
- **Hosting network players:** Frequently exchange innovations, both ways, with other units and its manufacturing staff communicates extensively with the other manufacturing managers in the network. They also are frequently hosting visitors from other units in the network.
- **Active network players:** Like hosting network players, but the level of communication centrality and the outflow of innovations are even higher and the major flow of visitors is in the opposite direction.

The different strategic role played by each type of plant is reflected also in the logistics organization of the plants characterized in terms of level of autonomy, level of resources, and investments.

However, the roles of the different facilities within their GMN are neither equal nor constant through time. It makes the analysis and management of a network much more complicated than only considering a plant as isolated (Feldmann and Olhager, 2010). That is why the understanding and deploying operationally the roles (and its drivers) could provide a systematic path for getting a much more efficient GMN.

While Ferdows' work provides a useful starting point for designing or restructuring the operations of plants within a GMN, there is little

Table 5.2 Classifications of strategic roles for multinational network plants

	Bartlett and Ghoshal (1989)	Jarrillo and Martinez (1990)	Ferdows (1997)	Veerecke et al. (2006)
Key dimensions for considerations	• Competence • Strategic alignment with national environment	• Competence • Integration level	• Competence • Location	• Intangible knowledge network • Physical, logistic networks
Strategic roles	• Implementer • Black hole • Contributor • Strategic leader	• Receptive • Active • Autonomous	• Offshore • Source • Server • Contributor • Outpost • Lead plant	• Isolated plants • Receivers • Hosting network players • Active network players

evidence of the application of the model beyond the work by Vereecke and Van Dierdonck (2002). This particular study discusses the application of Ferdows' model to the decision-making process of establishing and/ or acquiring a new production unit. However, it does not look into how the competencies of the production units within the existing network can be developed to enable the adoption of a higher level strategic role. In order to generate new insights into the application of the model further empirical research is required (Chakravarty, Ferdows, and Singhal, 1997). Furthermore, it is clear that more research is needed to understand how to coordinate the operations of individual production units within a network of manufacturing facilities (DuBois, Toyne, and Oliff, 1993; Shi and Gregory, 1998; Shi and Gregory, 2005). Models and techniques to aid practitioners formulating and developing operations strategy when designing or restructuring a Global Operations Network (GON) are lacking (Vereecke and Van Dierdonck, 2002) and the study areas are dispersed (Corti, Egaña, and Errasti, 2009; Laiho and Blomqvist, 2010), which results in difficulties to renew competences and capabilities of individual facilities (Teece, Pisano, and Shuen, 1997; Sweeney, Cousens, and Szwejczewski, 2007). Finally, there is limited research on the improvement programs and intrafirm practice transfer in multinational manufacturing enterprises with GONs (De Toni and Parusini, 2010).

The new paradigm in global operations strategy is, therefore, that there must be continuous reconfiguration of the manufacturing systems of a GMN in order to adapt its operations to be as efficient and effective as possible. The ability to quickly and effectively reconfigure the operations of the plants within the GMN is then a key source of competitive advantage.

Prior to developing the Akondia framework (explained in the next section) Mediavilla and Errasti had discussed in two previous research papers the usefulness of the Lean production based models for assessing the plant role suggested by Ferdows (Mediavilla and Errasti, 2010; Mediavilla, Errasti and Domingo, 2011). This analysis (carried out in the context of a multinational company with more than 40 plants worldwide) showed that Lean management models had significant limitations for operationalizing Ferdows' strategic plants roles; however, they were highly valuable for a cost-excellence or eventually a service-based strategy for improving a given plant role. Therefore, this chapter is structured as follows: (1) Explanation about the Akondia framework in order to be able to continuously identify a given plant role and prioritize its improvement; and (2) Overview regarding multisite improvement programs based on Lean management, which have probably the highest priority to improve industrial plants' roles.

The Akondia Framework

The proposed framework called *Akondia* aims to facilitate the practical application of Ferdows' model and to extend its application by systematically assessing and improving the competencies of a plant or a GMN based the plant role model (Mediavilla et al., 2012).

Depending on the analysis scope, the Akondia framework contributes to the continuous optimization and sustainability of: (1) GMN (by identifying strengths of each network unit and prioritizing the assignment/development of competences from network perspective) or (2) individual units (supporting plants to upgrade to a more attractive plant role).

Mediavilla and Errasti developed the Akondia framework as part of a team that included employees from a German-based multinational company and university lecturers. During the development phase, several key research works (e.g., Ferdows, 1997a; Vereecke and Van Dierdonck, 2002; Porter, 1985) and internal benchmarking methods utilized by the company (e.g., Lean-based production systems, purchasing excellence programs, supply chain standards, etc.) were utilized by the researchers. Parallel to the theoretical development of the framework, an assessment tool was defined in the form of a questionnaire.

The Akondia framework is divided into four stages. Table 5.3 provides an overview of each stage of the process.

Assessment of the Strategic Role of Plants: Stage 1, Stage 2, and Stage 3

Stage 1: Assess the Competencies of the Plant

The first stage of the framework aims to provide a strategic plant profile or competitive position as an output, which later could be compared to the generic roles defined by Ferdows in the second stage of the framework, and finally provides the basis for the gap analysis and improvement plan. The plant role, as originally defined by Ferdows, implicitly covers functions beyond purely production and supply chain activities, i.e., aspects within a GON that are not only part of a supply chain, but of the entire value chain. This is specially remarked on when introducing the concept of "Lead plant" (Ferdows, 1997a), a plant contributing to the company's strategy by developing capabilities as new processes, products, and technologies, and sharing these capabilities with other plants in the network. Therefore, the value chain concept defined by Porter (1985) was adopted as the basis for the plant role assessment proposed by the Akondia framework. Based on Porter's model, the

Table 5.3 Stages of the Akondia framework

Stage	Purpose	Who Is Involved	What to Do	Support Tools
Stage 1: Assess the competencies of the plant	To carry out an assessment of the plant's competencies based on Porter's value chain model	Researchers Business unit management team Local management of the plant	1. Interview business unit management team 2. Interview plant management team 3. Plot results in appropriate graphical format	Structured questionnaire based on Porter's value chain model
Stage 2: Develop generic profiles for all strategic roles	To identify the required and recommended competencies for each strategic role as defined by Ferdows and define their maturity level	Researchers Business unit management team	1. Delphi panel 2. Plot generic profiles in appropriate graphical format	Delphi panel results Competency assessment results
Stage 3: Define the plant role	To define the plant's role based on the analysis of the competency assessment	Researchers	1. Compare the generic plant role profiles with the results from the competency assessment 2. Define plant role based on closest match to generic plant's role profiles	Generic plant role profiles Competency assessment results
Stage 4: Define an improvement roadmap	To define an improvement roadmap to strengthening the competitive position of the GON or plant	Researchers Business unit management team Plant management team	1. Define improvement roadmap to strengthen current plant role 2. Define improvement roadmap to adopt a new plant role	Graphical prioritization matrix (competence level and influence level) Relevant improvement models and tools to support specific competence development needs

following six main fields of analysis were defined to carry out the plant role assessment:

1. Markets and customers
2. Suppliers
3. Internal operations
4. HR management
5. Technology management
6. Socio-political and regulatory issues

These six areas are mentioned by Ferdows in the description of the different generic plant roles.

In order to assess the competencies of the plant, a questionnaire was designed focusing on the six analysis fields mentioned above. The questionnaire includes 38 questions with each question focusing on a particular competence. Each competence is then evaluated under two dimensions: (1) level of influence of the plant to develop the competence (e.g., Is the plant able to select its strategic suppliers or is this centralized decision? Has the plant any influence in the new product technologies or is this carried out by another plant within the network?); and (2) current competence level.

The questionnaire then is to be used during structured interviews with individuals from the management team of the plant that is being assessed and from the top management team of the corresponding business unit (or headquarters). The aim of the latter is to provide an "external" view of a given plant in order to gain a better overview of the different plants within the network and enable richer comparisons. Furthermore, the headquarters are usually the decision-making agents for assigning, strengthening, or denying the competences for each plant or the plant's influence level on each competence.

The next step is to graphically plot the data gathered through the questionnaires. In the next example, it is possible to observe a real application of the Akondia framework in a globally operating corporation dedicated to the design, production, and distribution of consumer goods (six selected plants within the same business unit and located in different countries are shown). Figure 5.2 provides the detailed strategic profile showing the results of the 38 questions, while Figure 5.3 shows a summarized view of the questions based on the six main analysis fields.

The main output coming from this first stage is the strategic plant profile or competitive position. Apart from the overview of the strengths and weaknesses in each competence for an analyzed plant, the first graphical comparison of different plants provides an excellent overview of the competitive position of each plant within its GMN.

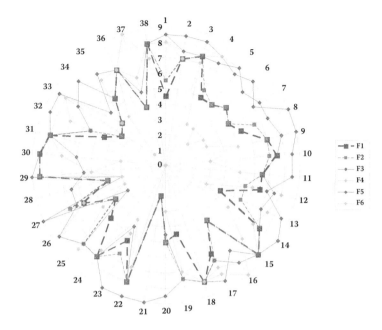

Figure 5.2 Strategic profile of six factories acting in the same business unit.

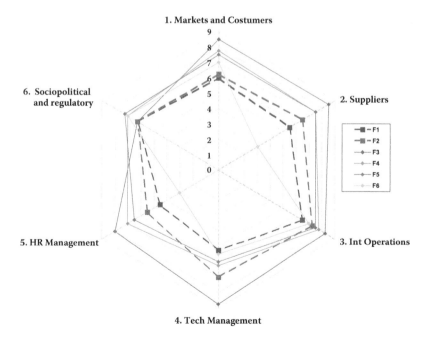

Figure 5.3 Consolidated view (mean per analysis field) of the strategic profile.

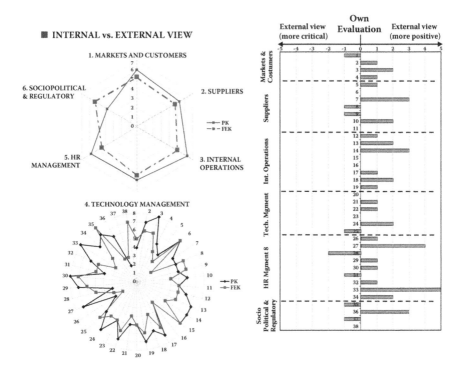

Figure 5.4 Competitive position (external versus internal evaluation).

An interesting complementary graphical plot is the comparison of the external and internal assessment, i.e., business unit or headquarters (HQ) assessment versus plant management assessment. This comparison is critical because it helps to check if everyone has a similar understanding of the competitive position of a plant or if there are correcting scaling differences in the answers given by different interviewees. Figure 5.4 illustrates a somewhat unusual example where the evaluation of the plant's management team has a much lower score than that carried out by the HQ.

Stage 2: Develop Generic Profiles for the Strategic Roles
The second step of the Akondia framework is an analytical phase in which, based on the competitive analysis carried out in Stage 1, a strategic plant role (as of Ferdows, 1997) is assigned to the plant. The step prior to doing this is to create generic profiles (specific to the GON under study) for each strategic plant role identified by Ferdows, i.e., "low-cost production" roles (offshore, source), "skills and knowledge" roles (outpost, lead), and "proximity to market" roles (server, contributor).

The authors developed generic profiles assigned to each of the Ferdows plant roles through the following steps:

1. Delphi panel with the HQ/business unit management team to relate each assessed plant with a particular strategic plant role.
2. Group the plants with the same strategic plant role.
3. Based on the evaluation done in the Stage 1, carry out a quantitative analysis and identification of the common competences that each plant of a given role shows.
4. In a second Delphi panel session, contrast the identified common competences and discuss/agree the "must" and "recommended" competences for each of the strategic plant roles.

The outputs from the above process were generic profiles for each strategic plant role that included the required competencies ("must" and "recommended") and their level. These generic profiles would then be used to compare each plant with the competitive assessment carried out in the previous stage. It is important to note that depending on a number of variables (e.g., sector, company size, product range, business unit) these generic profiles for each strategic plant role could be different (i.e., include different "must" and "recommended" competencies). Figure 5.5 and Figure 5.6 show two examples of generic strategic profiles developed for a certain case study.

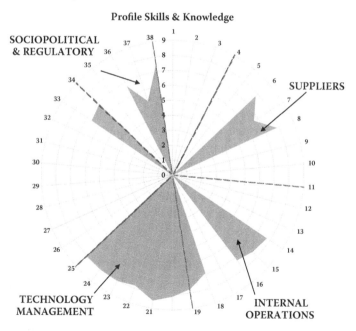

Profile Skills & Knowledge

Figure 5.5 Example of a generic strategic profile for a "lead" role showing the "must" competences.

Profile Low-Cost

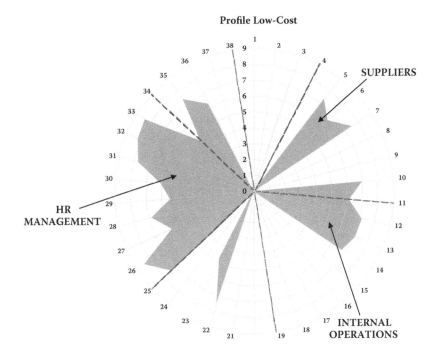

Figure 5.6 Example of a generic strategic profile for an "offshore" role showing the "must" competences.

Stage 3: Define the Plant's Role

The comparison of the strategic plant profile (executed in the stage 1) against the generic strategic profiles (stage 2) provides a quantitative and structured approach for defining the current plant role.

As already explained in stage 2, there are "must" and "recommended" competences for each generic strategic profile assigned to each of the Ferdows plant roles. These competences will be used to get the affinity level of a plant to a particular plant role by quantitatively comparing the current competence level (stage 1) to the required level for a generic strategic role (stage 2). This quantitative analysis provides an affinity level to each of the Ferdows plant roles.

For the case study presented in this chapter, the research team used a more pragmatic approach. The plants were first categorized into one of the three types of "strategic location advantage" defined by Ferdows (i.e., access to low-cost production, access to skills and knowledge, proximity to market). Once the dominating strategic location advantage is clearly defined, one of the two corresponding roles (e.g., "market orientation" strategic location advantage includes the "server" and "contributor" roles)

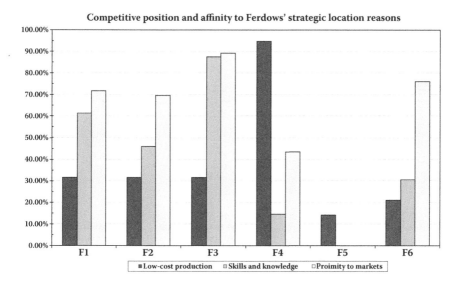

Figure 5.7 Affinity to Ferdows' model for six factories (based on competitive position).

is assigned depending on the competence level and influence (on competence) of the plant.

In Figure 5.7, the reader can observe the affinity level of six factories to Ferdows' model. Note that in these cases the initial affinity analysis has been done not to the six generic plant roles, but to the three strategic location advantage types. A similar affinity analysis can be carried out clustering the framework in the six generic plant roles.

In order to finalize the third stage, the most suitable graphical presentation is to summarize the entire affinity analysis by mapping each of the analyzed plants on Ferdows' model (Figure 5.8).

Roadmap Deployment for Upgrading the Strategic Role: Stage 4

Stage 4: Define an Improvement Roadmap for the Plant

The last step of the Akondia framework is focused on how to develop a different role for any analyzed plant (in most cases, this means adopting an even higher value-added role). In order to make this process systematic, it is highly recommended to prioritize the effort for achieving the new role status by the following two phases:

1. Strengthen the current plant role.
2. Develop improvement roadmap for achieving a new plant role.

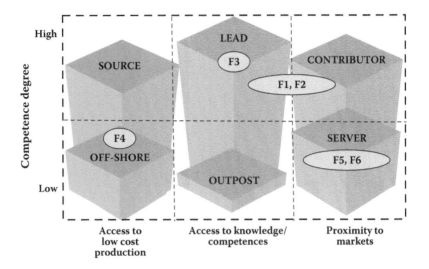

Figure 5.8 Plants and their Ferdows strategic role.

The Akondia framework has been developed in two levels of analysis: the first level should go through a "macro" perspective along the six main analysis fields, i.e., executed in the stages 1 and 2 of the framework. A second detailed and separated "micro" level approach per field can be utilized for the development of the improvement roadmap by adopting existing models and frameworks related to each analysis field (e.g., Lean Production for "Internal Operations," SCOR® (Supply Chain Operations Reference) model for "Markets and Customers," etc.). Only by clustering the improvement areas and defining development paths for improvement in given competences, will the plants be able to move into more valuable roles (Figure 5.9).

For strengthening the current plant role, it is necessary to identify the weakest competences. However, it is important to remember that the questionnaire allowed assessing two aspects of each plant: the competence level and the influence level of the plant in that competence. It is logical to assume that the higher the influence level on a given competence, the easier to improve its level. The Akondia framework proposes first strengthening of the current strategic role by analyzing the prioritization matrix in Figure 5.10.

The prioritization matrix shows graphically each of the 38 assessed competences in the questionnaire: the *y*-axis shows the level of each competence, while the *x*-axis provides the influence of a given plant on the development of that particular competence. Another graphical alternative for a quick identification of the fields requiring improvement is shown in

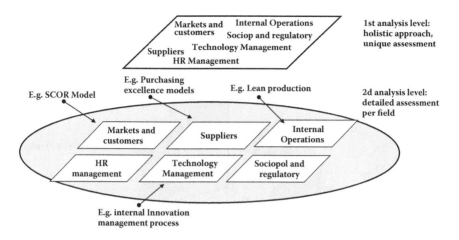

Figure 5.9 Analysis perspectives of the Akondia framework.

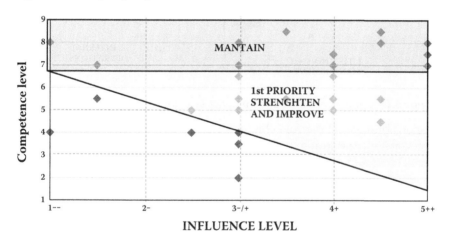

Figure 5.10 Priorization matrix based on questionnaire scoring.

Figure 5.11. Note that the highest influence level is scored as 5 while the scale for the competences level goes up to 9.

After strengthening the current competitive position by a focused improvement on the competences where the influence is high, any plant could be suitable for developing a strategic role roadmap. Using the generic competitive positions (or strategic profile) for the strategic roles of Ferdows, a step-by-step, middle- to long-term roadmap can be deployed.

We recommend a deployment roadmap based on the six analysis fields, balancing the current competence level, the influence level, and the effort required (e.g., providing more influence to a facility by giving it new

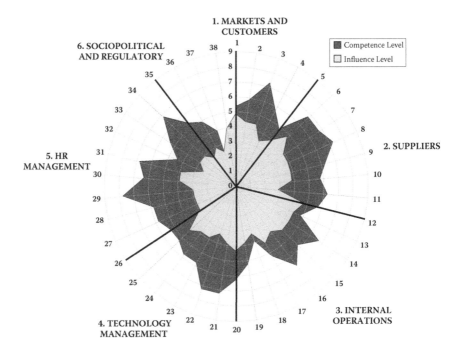

Figure 5.11 Competence level and influence level for a facility.

responsibilities). These role changes, in fact, could imply organizational decisions or reassignment of responsibilities. Any improvement roadmap per field should have a second-level analysis and detailed deployment at the operational level.

From the practical experience of applying the "Akondia framework" in the industry, we are able to summarize a detailed overview of the "must" and "recommended" competences per Ferdows' factory role. Refer to Table 5.4 for a detailed overview, which shows the mentioned "must" and "recommended" competences for each strategic site reasons. Note that a similar matrix can be shown for the influence level per competence.

Multisite Improvement Programs Based on Lean Management

Multinational manufacturing companies are increasingly implementing lasting corporate improvement programs. These systems are often labeled "X Production Systems" (XPS), where the X denotes the company's name (Netland, 2012). The aim is to improve operational efficiency by sharing best operational practices and fostering continuous improvement among

Table 5.4 "Must" and "recommended" competences per Ferdows' factory role

Strategic Site Reason	"Must" Competences	"Recommended" Competences
Access to low-cost production OFFSHORE	Availability of low-cost blue collars Availability of low-cost engineers	Access to a cost-competitive network of suppliers Capacity of having a stable supply process from low-cost suppliers Efficiency of utilization of the workforce Efficiency of utilization of the machines/installations
Access to low-cost production SOURCE	Access to a cost-competitive network of suppliers Access to reliable and flexible "order–delivery" processes from the network of suppliers Capacity of having a stable supply process from low-cost countries (LCC) "Purchase productivity" (administrative price ratio)	Integration/response with customer service of main markets "Purchasing productivity" (technical–material ratio) Capacity of successful collaboration/integration with suppliers Capacity of robust supplier administration contract management (currencies, delivery terms, penalties, etc.), financial risks monitoring Efficiency of utilization of the workforce Efficiency of utilization of the machines/installations Capacity of maintaining a reliable delivery fulfillment Flexibility to adapt the production to changes in the demand and special occasions (seasonality, special promotions, etc.) Commitment/motivation of the employees Capacity of maintaining low absenteeism rate Attachment of the plant to complaint behavior

(continued)

Table 5.4 "Must" and "recommended" competences per Ferdows' factory role (continued)

Strategic Site Reason	"Must" Competences	"Recommended" Competences
Access to skills and knowledge **OUTPOST**	Proximity to a knowledge/technology networks source Availability of high-skilled blue collars Availability of high-skilled engineers	Integration/response with customer service of main markets Capacity of innovating in utilization of new process technologies Capacity of improving existing manufacturing processes (labor rationalization, Q-improvement) Capacity of implementing existing tools/techniques/methods (best practice) or develop them Availability for developing other factories/subsidiaries
Access to skills and knowledge **LEAD**	Integration/response with customer service of main markets "Purchase productivity" (administrative–price ratio) "Purchasing productivity" (technical–material ratio) Capacity of successful collaboration/integration with suppliers Efficiency of utilization of the machines/installations Efficient utilization of energy and appropriate waste management Capacity of serving as a pilot for new introductions	Capacity of offering value-added services to customers aligned with main subsidiaries of strategic markets Capacity of robust supplier administration: contract management, financial risks monitoring Efficiency of utilization of the workforce Capacity of maintaining low number of incidents and accidents Attachment of the plant to complaint behavior

(continued)

Table 5.4 "Must" and "recommended" competences per Ferdows' factory role (continued)

Strategic Site Reason	"Must" Competences	"Recommended" Competences
	Proximity to a knowledge/technology networks source	
	Capacity of innovating in product technologies	
	Capacity of minor adaptations/face lifting of existing products/platform	
	Capacity of innovating in utilization of new process technologies	
	Capacity of improving existing manufacturing processes	
	Capacity of implementing existing tools/techniques/methods (best practice) or developing them	
	Availability of highly skilled blue collars	
	Availability of highly skilled engineers	
	Availability for developing other factories/subsidiaries	
Proximity to markets SERVER	Proximity to strategic markets	Integration/response with customer service of main markets
	Robust phase-in/phase-out fulfillment	Capacity of having a stable supply process from low-cost suppliers
	Access to a cost-competitive network of suppliers	Efficiency of utilization of the workforce
	Incentives for investments, innovation, or keeping the operations	Efficiency of utilization of the machines/installations
		Flexibility to adapt the production to changes in the demand and special occasions
		Stability in the macro economical situation

(continued)

Table 5.4 "Must" and "recommended" competences per Ferdows' factory role (continued)

Strategic Site Reason	"Must" Competences	"Recommended" Competences
Proximity to markets CONTRIBUTOR	Proximity to strategic markets Robust phase-in/phase-out fulfillment Integration/response with customer service of main markets Access to a cost-competitive network of suppliers Access to reliable and flexible "order-delivery" processes from the network of suppliers "Purchase productivity" (administrative–price ratio) "Purchasing productivity" (technical–material ratio) Capacity of successful collaboration/integration with suppliers Capacity of managing a reliable delivery fulfillment Flexibility to adapt the production to changes in the demand and special occasions Capacity of innovating in utilization of new process technologies Capacity of improving existing manufacturing processes Capacity of implementing existing tools/techniques/methods (best-practice) or develop them Availability of highly skilled blue collars Availability of highly skilled engineers Incentives for investments, innovation, or keeping the operations	Capacity of offering value-added services to customers aligned with main subsidiaries of strategic markets Capacity of having a stable supply process from low-cost suppliers Capacity of robust supplier administration contract management, financial risks monitoring Efficiency of utilization of the workforce Efficiency of utilization of the machines/installations Efficiency of utilization of energy and appropriate waste management Capacity of serving as a pilot for new introductions Capacity of minor adaptations/face lifting of existing products/platform (extend/reduce features) Capacity of maintaining low absenteeism rate Availability for developing other factories/subsidiaries Stability in the macro economical situation Stability of the labor law framework Attachment of the plant to complaint behavior

all subsidiaries. These programs are heavily inspired from the Lean Management principles and deployed based on the Toyota Production System. The most advanced improvement programs are integrating detailed value chain activities evaluations (Mediavilla and Errasti, 2010; Mediavilla, Errasti, and Domingo, 2011).

The strategic objective of XPSs is often abbreviated to C, Q, S, F, I, R, S_2, P, and E. These letters stand for the traditional competitive priorities measurement key performance indicators (Gobbo, 2007): cost (C), quality (Q), speed (S), flexibility (F), innovation (I), and reliability (R). In addition, safety (S_2), people development (P), and environmental performance (E) occur in practice. Thus, XPSs aim to improve the competitiveness of multinational companies on several performance parameters and represent a holistic approach to how multinational companies can do process improvement.

XPS Principles

The XPSs are usually created as a consequence of the general belief in transferability of a management system across borders, regardless of the cultural differences (Harbison and Myers, 1989). They are often associated with "best practices" that are believed to be applicable across nations (Koontz, 1969; Kono, 1982). The most common "best practice" model is the Toyota Production System (TPS) and its descendant *Lean thinking*. Liker's book, *The Toyota Way* (McGraw-Hill, 2004), offers the description of the Toyota Production System shown in Figure 5.12.

Multinational companies tailor the Lean thinking to their specific environments and needs. XPSs are usually structured in principles, methods, and tools in a hierarchical and well-defined structure.

- Principles: What do we want to achieve (objectives)?
- Methods: How do we want to reach the objectives?
- Tools: Which tools or techniques can we use to achieve the objectives?

A close relationship between the main principles of modern XPSs and the Toyota Production System has been established in literature (Neuhaus, 2009; Netland, 2012). Netland analyzes 30 XPSs and finds that the 12 most common XPS principles are the ones shown in Figure 5.13.

The Management of XPSs

More and more companies in more and more industries reap the benefits of a transition to Lean thinking through the development and deployment of an XPS. The successful implementation of an XPS can increase competitiveness through cost reductions, better quality, and increased

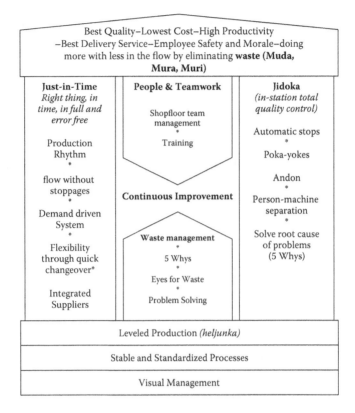

Figure 5.12 Framework structure of the Toyota Production Systems principles and methods/tools. (Adapted from Liker, J. K. (2004) *The Toyota way: 14 management principles from the world's greatest manufacturer.* McGraw-Hill, New York.)

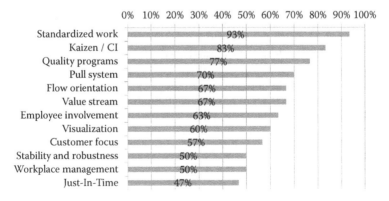

Figure 5.13 The most common XPS principles in practice. (From Netland, T. H. (2012) Exploring the phenomenon of company-specific Production Systems: One-best-way or own-best-way? *International Journal of Production Research.*)

sales. Hence, it should be considered as a necessary strategy by most multinational companies. However, one should be aware that the true competitive advantage is not the system per se, but tied to the process and culture. An XPS is *not* a panacea for competitiveness. Companies should not put a lot of effort into developing the XPS system and structure and simultaneously ignore the cultural-dependant implementation. The true source for sustainable competitive advantage is to be found in the organizational process of implementing and applying the XPS, and not in its content (Netland and Aspelund, 2013).

To implement the XPSs in the subsidiaries, multinational companies establish organizational XPS structures. Often an XPS Program Office is established at the corporate headquarter. The Program Office is responsible for the development and implementation of the XPS. Often it oversees the XPS documentation, assesses the XPS maturity in subsidiaries (as described in the next section), helps spread best practices among subsidiaries, and gives general support to XPS implementation. The XPS Program Office coordinates implementation with a global XPS organization; typically each business unit has a global XPS director and each facility has a XPS coordinator. Figure 5.15 illustrates a typical XPS organizational structure in a multinational company.

Succeeding with an XPS starts and ends with management commitment. Countless studies of critical success factors (CSF) in the scholarly literature establish that one will never succeed with a change program without a profound management commitment. The Jotun case shows how a growing multinational company works strategically with management commitment.

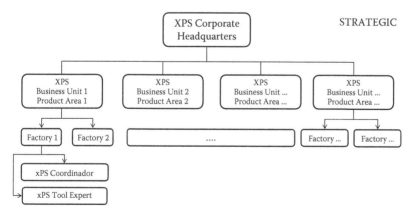

Figure 5.15 Example of an XPS organization within a multinational company.

CASE: The Jotun Operations Academy (Aa and Anthonsen, 2011)

The Jotun Group is one of the world's leading manufacturers of paints, coatings, and powder coatings. Jotun is headquartered in Sandefjord, Norway, and has 40 plants and 8600 employees worldwide. Jotun develops, produces, and sells various paint systems and products to protect and decorate surfaces in the residential, shipping, and industrial markets. Jotun enjoys a rapid growth due to its world-class products and services. The need for increased capacity during the next decade will be met with a combination of greenfield investments, major expansion projects in current facilities and improvement work based on the Jotun Operations System (JOS).

In order to accommodate the growth and improve operational performance in all plants, Jotun has established arenas for managers and employees to come together and share best practices. One of these arenas is Jotun Operations Academy (JOA), which is an extensive competence development programme. The JOA is hence a key element in the worldwide implementation of Jotun's XPS: the JOS.

The JOA contributes with sense-making and learning opportunities to employees and is believed to foster a deeply rooted culture change. It is an important arena to transfer best operational practices between Jotun plants. Jotun's own managers and trainers who embody the four company values—loyalty, care, respect and boldness—transfer values and practices between the different Jotun entities. Jotun believes in a values-based approach in all parts of the organization where mutual respect and cooperation between employee and manager is the road to success.

In order to accommodate the growth and improve operational performance in all plants, Jotun has established arenas for managers and employees to come together and share best practices.

Figure 5.16 Jotun's products

Benchmarking and Assessment Models

Because the XPS aims to implement best practices in subsidiaries, *benchmarking* becomes a very useful tool for keeping track of development and spreading practices. Benchmarking is the process of comparing one factory's processes with other factories' processes. Because the XPS processes can be seen as "globally valid best practices" (Koontz, 1969; Kono, 1992), a standard audit scheme easily can be developed and used for assessment across plants.

Many multinationals use assessment models strategically to push the implementation of XPSs in plants. The idea is that when it is measured and requested it will be done. Often, detailed assessment documentation and procedures are developed and carried out by onsite visits in all plants worldwide. The following case gives a brief example of how this is done in a global company. Alternatively, the assessments can be carried out as self-assessments. The objective is to establish the current maturity stage and point out the next steps toward operational excellence. In this sense, it works as a guideline for XPS implementation.

CASE: German White Goods Production System Assessment Model (Mediavilla and Errasti, 2010)

This white goods company is a global company with more than 40 factories in 13 companies in Europe, Latin America, North American, and Asia, which employs 45,000 worldwide.

As part of an effort to increase competitiveness on a global basis, the company has established its own XPS. The aim is to improve the productivity of all the plants in the network. An important part of the production system is the assessments that the company is carrying out in all plants. The assessment is built up around two questionnaires: one measures the *implementation grade* of the XPS in the plants, while the other measures the *performance maturity grade* (Figure 5.17).

There are a total number of 400 questions in the *implementation* questionnaire and 150 in the maturity questionnaire. A scoring system allows a maximum score of 1000 points for each questionnaire. The scores are merged into overall scores that define the classification of the plants as either A level (>850 points), B level (>700 points), or C plants (>500 points). For the "degree of implementation," a criterion regarding the wideness of the implementation is followed:

- 0: There is nothing planned or implemented
- 1: The XPS is planned or implemented
- 2: The XPS is implemented in a pilot area
- 3: The XPS is implemented in >50% of the plant
- 4: The XPS is fully implemented throughout the plant

There are a total number of 150 questions in the *maturity* questionnaire. Also, a total number of 1000 points is max. For the maturity questionnaire, all questions have quantitative key performance indicators associated with it that serves as the basis for scoring. The maturity questionnaire suggests the following scale:

- <200 points: Plants with low efficiency in the implementation
- 200–400 points: Advanced factory situation and good use of method
- 400–700 points: Excellent efficiency in the use of the model
- >700 points: Ideal state of factories

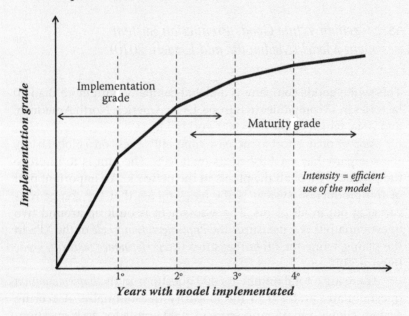

Figure 5.17 Implementation grade and maturity grade relation.

CASE: Natra

With more than 50 years of history, Natra is today a leading company in Europe specializing on chocolate products for the private label brand as well as cocoa derivatives for the food industry.

Natra produces candy bars, chocolates and Belgian specialties, tablets, and chocolate spreads. The company has five production centers to offer one of the most extensive catalogs available in Europe.

The strategic positioning in the American and Asian market as well as the growth in the European market is being supported by the commitment to the research and innovation of new recipes, packaging and tailor-made solutions.

The global production network has also adopted the own best way approach to gain Operational Excellence called Natra Production System.

This system is composed of principles, values and behaviors at work, Operating Model and Processes, Goals and Measurement System, and Methods and Tools.

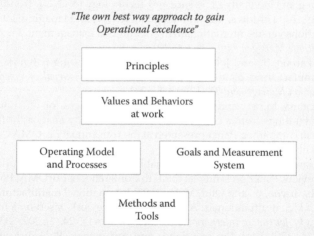

"*The own best way approach to gain Operational excellence*"

Principles: rules and beliefs governing individual and teams behaviour derived from company's operations strategy, but also management, operational effectiveness and efficiency philosophies assumptions.

Values/behaviors: ways of acting in everyday situation at work that aids reaching the vision and goals

Operating Model and Processes: The organization functions and tasks for Shopfloor management to carry out the define map of processes of the manufacturing Value Chain with integrated information systems

Goals and Measurement System: performance objectives (Cost, Quality, Delivery, Innovation, Safety) derived from the strategic planning and the annual budget.
What do we want to achieve?

Methods: the implementation/improvement process description
How do we want to reach the objective/goal?

Tools: the specific techniques within a method that aids the method implementation
Which means do we use to achieve objectives?

References

Aa, O. A., and Anthonsen, H. (2011) Management of best practices in multinational companies: A comparative case study concerning implementation of operations best practices in two subsidiaries of the Jotun Group. Master's thesis, Norwegian University of Science and Technology (NTNU), Trondheim.

Chakravarty, A., Ferdows, K., and Singhal, K. (1997) Managing international operations versus internationalizing operations management. *Production and Operations Management* 6 (2): 100–101.

Cheng, Y., Farooq, S., and Johansen, J. (2011) Manufacturing network evolution: A manufacturing plant perspective. *International Journal of Operations and Production Management* 31 (12): 1311–1331.

Corti, D., Egaña, M. M., and Errasti, A. (2009) Challenges for off-shored operations: Findings from a comparative multi-case study analysis of Italian and Spanish companies. Paper presented at the 16th annual EurOMA Conference, Gothenburg, Sweden.

De Toni, A., and Parussini, M. (2010) International manufacturing networks: A literature review. Paper presented at the 17th Conference of EurOMA, Porto, Spain.

Dubois, F. L., Toyne, B., and Oliff, M. D. (1993) International manufacturing strategies of U.S. multinationals: A conceptual framework based on a four-industry study. *Journal of International Business Studies* Q2 24 (2): 307–333.

Feldmann, A. (2011) Studies in science and technology. PhD diss., Linköping University, Sweden, no. 1380.

Ferdows, K. (1997) Making the most of foreign factories. *Harvard Business Review* March–April: 73–88.

Kono, T. (ed.) (1992) Japanese management philosophy: Can it be exported? In *Strategic management in Japanese companies* (pp. 11–24). Pergamon Press, Oxford, U.K.

Koontz, H. (1969) A model for analyzing the universality and transferability of management. *The Academy of Management Journal* 12 (4): 415–429.

Laiho, A., and Blomqvist, M. (2010) International manufacturing networks: A literature review. Paper presented at the 17th Conference of EurOMA, Porto, Spain.

Liker, J. K. (2004) *The Toyota way: 14 management principles from the world's greatest manufacturer*. McGraw-Hill, New York.

Mediavilla, M., and Errasti, A. (2010) Framework for assessing the current strategic plant role and deploying a roadmap for its upgrading. An empirical study within a global operations network. Paper presented at the APMS 2010 Conference, Cuomo, Italy.

Netland, T. H. (2012) Exploring the phenomenon of company-specific production systems: One-best-way or own-best-way? *International Journal of Production Research*.

Netland, T. H., and A. Aspelund (2013) Company-specific production systems and competitive advantage: A resource-based view on the Volvo production system. *International Journal of Operations & Production Management*.

Neuhaus, R. (2009) Produktionssysteme in deutshen Unternehmen—Hintergründe, Nutzen und Kernelemente. Fachzeitschrift Industrial Engineering, *REFA Bundesverband*, pp. 24–29.

Porter, M. E. (1985) *The competitive advantage: Creating and sustaining superior performance*. Free Press, New York.

Shi, Y., and Gregory, M.J. (1998) International manufacturing networks: To develop global competitive capabilities. *Journal of Operations Management* 16: 195–214.

Shi, Y., and Gregory, M. (2005) Emergence of global manufacturing virtual networks and establishment of new manufacturing infrastructure for faster innovation and firm growth. *Production Planning & Control* 16 (6): 621–631

Sweeney, M., Cousens, A., and Szwejczewski, M. (2007) International manufacturing networks design: A proposed methodology. Paper presented at the 2007 EurOMA Conference, Ankara.

Teece, D. J., Pisano, G., and Shuen, A. (1997) Dynamic capabilities and strategic management. *Strategic Management Journal* 18 (7): 509–533.

Vereecke, A., and Van Dierdonck, R. (2002) The strategic role of the plant: Testing Ferdows' model. *International Journal of Operations & Production Management* 22 (5).

Further Readings

Abo, T. (1994) *Hybrid factory: The Japanese production system in the United States*. Oxford University Press, New York.

Agrawal, V., Farrell, D., and Remes, J. (2003) Offshoring and beyond. *McKinsey Quarterly* 3: 24–35.

Bartlett, C. A., and Ghoshal, S. (1989) *Managing across borders. The transnational solution*. Boston: Harvard Business School Press.

Beechler, S., and Yang, J. Z. (1994) The transfer of Japanese-style management to American subsidiaries: Contingencies, constraints, and competencies. *Journal of International Business Studies* 25 (3): 467–491.

Corti, D., Pozzetti, A., Zorzini, M. (2006) Production relocation of Italian companies in Romania: An empirical analysis. Paper presented at the proceedings of the EurOMA Conference, Glasgow, June 18–21, 1: 21–30.

De Meyer, A., and Vereecke, A. (1996) International operations. In *International encyclopedia of business and management*, ed. M. Werner. Routledge, London.

Feldman, A. and Olhager, J. (2010) Linking networks and plant roles: The impact of changing a plant role. Proceedings of the 17th EurOMA Conference, Porto, Portugal, June 6–9.

Flaherty, M. T. (1986) Coordinating international manufacturing and technology. In *Competition in global industries*. M. E. Porter (ed.) Boston: Harvard Business School Press.

Frick, A., and Laugen, B. (eds.). (2011) Advances in production management systems. Value networks: Innovation, technologies, and management IFIP WG 5.7. International Conference, APMS 11, Stavanger, Norway, September 26-28. Revised Selected papers Series: IFIP Advances in information and Communication Technology, 384: 354–363.

Gobbo, J. (2007) Inter-firm network: A methodological approach for operations strategy. Proceedings of the 14th EurOMA Conference, Ankara, Turkey, June 17–20.

Jarillo, J. C., and Martinez, J. L. (1990) Different roles for subsidiaries: The case of multinational corporations in Spain. Strategic Management Journal 11(7): 501–512.

Kumon, H., and Abo, T. (2004) *The hybrid factory in Europe: The Japanese management and production system transferred.* Antony Rowe Ltd., London.

Koren Y., Heisel, U., Jovane, F., Moriwaki, T., Pritschow, G., Ulsoy, G., and Van Brussel, H. (1999) Reconfigurable manufacturing systems. *CIRP Annals, College International de Recherches pour la Production* 48 (2) : 527–540.

Mediavilla, M. Errasti, A., and Domingo, R. (2011) Modelo para la evaluación y mejora del rol estratégico de plantas productivas: Caso de una red global de operaciones. *Dyna Ingeneria e Industria,* Agosto 86 (4): 405–412.

Mediavilla, M. Errasti, A., Domingo, R., and Martinez, S. (2012) Value chain based framework for assessing the Ferdows' strategic plant role: An empirical study. *APMS* 369–378.

Mehrabi, M., Ulsoy, A., and Koren, Y. (2000) Reconfigurable manufacturing systems: Key to future manufacturing. *Journal of Intelligent Manufacturing* 11 (4): 403–419.

Meijboom, B., and Voordijk, H. (2003) Internacional operations and location decisions: A firm level approach. *Voor Economische En Sociale Geografie* 94 (4): 463–476.

Mintzberg, H., Lampel, J., Quinn, J. B., and Ghoshal, S. (1996) *The strategy process,* 4th ed. Prentice Hall, Hemel Hempstead, U.K.

Netland, T. (2011) Improvement programs in multinational manufacturing enterprises: A proposed theoretical framework and literature review. Paper presented at the EurOMA 2011 Conference, Cambridge, U.K.

Rudberg, M. (2004) Linking competitive priorities and manufacturing networks: A manufacturing strategy perspective. *International Journal of Manufacturing Technology and Management* 6 (1–2): 55–80.

Slack, N., and Lewis, M. (2002) *Operations strategy,* 2nd ed. Prentice Hall, Upper Saddle River, NJ.

Vereecke, A., Van Dierfonck, R., and De Meyer, A. (2006) A typology of plants in global manufacturing networks. *Management Science* 52 (11): 1737–1750.

Vokurka, R. J., and Davis, R. A. (2004) Manufacturing strategic facility types. *Industrial Management and Data Systems* 104 (6): 490–504.

Volvo Group. (2010) Annual report.

Womack, J., Jones, D., and Roos, D. (1990) *The machine that changed the world.* Massachusetts Institute of Technology (MIT). MacMillan, New York.

Yokozawa, K., de Bruijn, E. J., and Steenhuis, H.-J. (2007) A conceptual model for the international transfer of the Japanese management systems. Paper presented at the 14th International Annual Conference of the European Operations Management Association (EurOMA), Ankara, Turkey.

chapter 6

Factory and Facility Material Flow and Equipment Design

Sandra Martínez and Ander Errasti

Starting a great project takes courage. Finishing a great project takes perseverance

Contents

Introduction

In this chapter, we discuss:

- Facility classification based on volume/product and process
- Facility material flow and equipment design considerations in an offshore facility
- Facility planning process and layout design (workstations and areas)

Facilities Design Considerations

Facilities are critical nodes of a global manufacturing network from the perspective of the supply chain system. If we want to implement the most efficient principles, such as the demand-driven supply chain (Christopher, 2005), the design of facilities should consider management principles that are oriented toward increasing customer satisfaction, reducing total cost, and increasing the return of capital (Figure 6.1).

Once the facility location (see Chapter 4) and strategic role (see Chapter 5) are settled, the next step is the facility's design.

Tompkins et al. (2010) and Muther (1981) state that the elements of a facility consist of the facility's systems, the layout, and the handling systems.

The facility systems consist of the structural system, the enclosure systems, the lighting, electrical and communication systems, etc. (Figure 6.2).

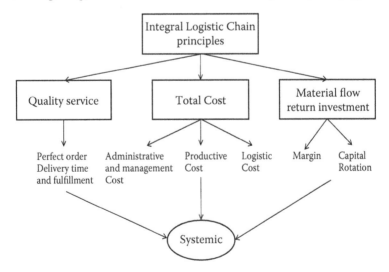

Figure 6.1 Integral logistic chain principles. (From Errasti, A. (2006) KATAIA: Modelo para el análisis y despliegue de la estrategia logística y productiva. PhD diss., University of Navarro, Tecnun, San Sebastian, Spain.)

Figure 6.2 Example of the structural system of a facility.

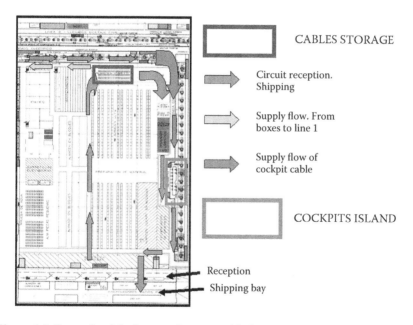

Figure 6.3 Example of the layout of an assembly line.

The layout consists of equipment and machinery distributed in the production and logistic areas, support areas, etc., within the building (Figure 6.3).

This layout is greatly influenced by the consideration and choice of what should move and what should be fixed within the building where the production is going to take place. There are several **combinations of moving and fixed materials, machine, and personnel alternatives, and**

Table 6.1 Different configurations of materials, machines, and personnel alternatives when designing a facility's layout

Materials	Machines	Personnel	Example
Fix	Move	Move	Shipyard
Move	Fix	Fix	Assembly line
Move	Move	Fix	AS/RS
Move	Fix	Move	Nagare assembly line

each one determines the materials flow performance and the subsequent handling systems.

For the same facility problem, there could be different alternatives for which different concepts with different handling systems are applicable, as seen in Table 6.1.

Another example in warehousing could be the parts-to-picker and picker-to-parts alternatives when designing a picking system (Figure 6.4 and Figure 6.5).

What Comes First, the Layout or the Handling Systems?

Even if the manufacturing process reveals the sequence of activities to be performed, it is not appropriate to address the material handling solutions after the layout macro configuration is finished.

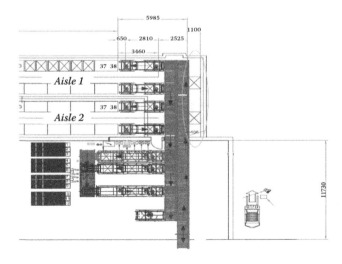

Figure 6.4 Example of a parts-to-picker AS/RS (Automated Storage/Retrieval Systems) with two workstations and a put-to-light system. (From ULMA Handling, San Sebastian, Spain. With permission.)

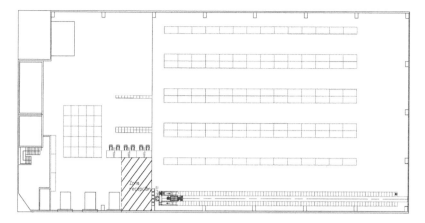

Figure 6.5 Example of a distribution center with conventional rack systems and picker-to-parts picking and a parts-to-picker miniload. (From ULMA Handling, San Sebastian, Spain. With permission.)

For instance, the material handling unit between machines and production areas and production lot size could determine two radically different alternatives, each influenced by different management principles (**Concept A and Concept B**) when applying handling efficiency considerations:

- **Concept A**: Cellular-balanced multiproduct manufacturing responding to leveled demand and short lead time with more short distance regular movements of one unit handling all units or one-piece flow between workstations. A **Lean production approach** would emphasize a one-piece flow paradigm, eliminating transfer of lot size and balancing the workstations in a U-shape layout to maximize employee utilization.
- **Concept B**: Line production balancing demand volume and complexity in the system constraints. This concept maximizes throughput and subordinates transfer lot size scheduling to bottlenecks. The work in process is subordinated to them, minimizing movements and optimizing transferring load unit between machines. This eliminates the need to reduce distance between machines. In contrast, a **theory constraint approach** that considers the difficulty of balancing the capacity and setup times would emphasize for specific mix products what the constraint would be and propose that a buffer/stock be decoupled from the upstream flow, and it would not necessarily propose cellular manufacturing, but rather line manufacturing, which optimizes lot size for the constraint and specific load unit transfer size.

This chapter will take an in-depth look at layout and handling systems through consideration of the aspects of material flow and equipment design. (This book is not necessarily for managers who are interested in structural systems design.)

Offshore Facility Design Considerations

When talking about offshoring, different options have been put forth in the literature; these options depend on aspects of spatial dimension and ownership. Nevertheless, this book is focused on international issues (Table 6.2).

Considering the dynamic point of view of the global markets, the design of new offshore static facilities would not be suitable for the entire life cycle of the plant.

That means that very few companies will be able to retain their facility or layout without severely damaging their competitive position in the marketplace. Thus, **the design should consider not only efficiency performance parameters, but also manufacturing reconfiguration and adaptive paradigms** (see Chapter 5).

In conclusion, **offshore facilities material flow and equipment design has to consider efficiency parameters along with the ability to continually update operations through relayout** and rearrangement.

In addition, managers generally believe that a facility and management system can be replicated or "copy-pasted" in any other site in the world. Nevertheless, that is not true because recent research (Errasti, 2009) highlighted that process and management **system modification and adaptation to local characteristics,** such as labor cost, equipment maintainability, product demand variety and volumes compared to the matrix facility, and supplier local network, **is a need that is not always considered by managers.**

Table 6.2 Different offshoring options depending on the spatial dimension and ownership

		Ownership	
		Internal	External
Spatial dimension	National	Domestic relocation	Domestic outsourcing
	International	Foreign relocation	Foreign outsourcing

Facility Material Flow and Equipment Design Considerations

The general aims of a facility's material flow and equipment design include:

- Improve customer satisfaction by responding to order delivery process needs and general customer service complaints and needs.
- Increase return of assets (ROA) by reducing material handling, maximizing inventory turns, and maximizing employee efficiency through workstation and production area design.
- Be adaptable and reconfigurable for future needs.
- Effectively use equipment, space, and energy through shop floor management systems to increase equipment availability (Takahashi and Takahashi, 1990) and work on process reduction (Womack and Jones, 2003) and energy waste reduction.
- Increase productivity through visual quality and flow production control (Hirano, 1998) and shop floor management systems in order to work on standardizing processes and involving workers in the continuous improvement of operational performance (Suzaki, 1993).
- Provide for employee safety and meet ergonomic as well as environmental requirements.

When designing or redesigning the facility, the company's value chain and extended supply chain functions have a significant impact (see Chapter 5) (Table 6.3).

Facility Planning Process

Previously, we have seen that the layout and the handling system should be designed simultaneously (Tompkins et al., 2010), and that is why once the basic requirements for each production areas have been collected, a battery of possible alternative block layouts should be proposed before a detailed layout is developed.

The facilities planning process for manufacturing and assembly facilities can be as follows (Tompkins et al., 2010; Muther, 1981):

- Define the products to be manufactured and assembled.
- Specify the required manufacturing and/or assembly processes and related activities.
- Determine the interrelationships among activities.
- Determine the space requirements for all activities.
- Generate, evaluate, and select a facility plan.

Table 6.3 Value chain functions and impact when redesigning the layout

Value Chain Functions	Feature	Variables or Drivers
Demand management	Product mix, product demand	Demand and variety responsiveness or flexibility
Product development	Product complexity, modularity, standardization, and design stability	Product value proposition
Manufacturing network design	Integrated versus fragmented manufacturing Focus versus redundant factories	Global network configuration
Process development	Plant sized, machineries technology, automation types and level, production control systems	Process know-how
Production planning and scheduling	Lot size, scheduling priorities, fix production planning horizons, staffing levels	Operational know-how
Service policy and order delivery strategy	Make to stock, make to order, assembly to order delivery strategy Storage needs and material flow local–global pull–push logic	Time responsiveness
Supplier development	Source location, type of source (components, subassemblies, etc.), and supplier value-added services (co-investment, co-design, logistic JIT flows, replenishment systems, etc.)	Extended supply chain development
Human resource management	Labor recruitment policies, labor skills, and ability formation	Resource management policies

- Implement the facility plan.
- Maintain and adapt the facility plan.
- Update strategic role and the facility, update the products to be manufactured and/or assembled, and redefine the layout (see Chapter 5).

Once the product, process, and schedule design decisions have been made, the facilities planning team needs to generate and evaluate the layout and handling alternatives. For this purpose, some authors (e.g., Tompkins et al., 2010) propose seven management and planning tools.

Specify the Required Manufacturing and/or Assembly Processes

Once the product variety and volume are known, the manufacturing process has to be designed. For this purpose, the operation/flow process chart is a useful graphic tool for showing the number of operations (production, transportation, storage), inspections, and operations times.

This chart could be interpreted as a network (Muther, 1981) or a production process "flow," such as a river flowing into the sea. Just as river collects water from rain and from other streams, it may start running downhill right away. Similarly, components and raw materials are the river's beginnings.

Water runs in channels over the land and we call them **streams** and **rivers**. The manufacturing processes are the river's middle point. If the water is contained and slow, we call them **lakes** and **ponds**. Similarly, this happen with the stock in process because the stock could be buffered between manufacturing or assembly areas.

Leaving the high elevation of mountains and hills and entering the flat plains, the river slows down and changes into a main stream. These are the manufacturing process's main assembly or manufacturing lines or cells.

Finally, the river flows into another large body of water, such as an ocean, bay, or lake. Similarly, the manufacturing ends in the dispatching or delivery zone (Figure 6.6).

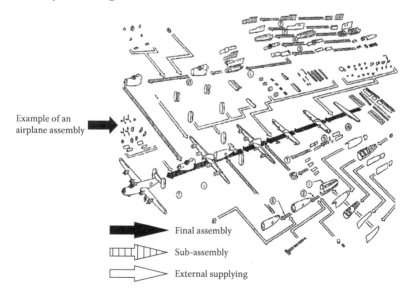

Example of an
airplane assembly

Final assembly

Sub-assembly

External supplying

Figure 6.6 Example of an airplane assembly process. (Adapted from Muther, R. (1981) *Distribución en planta*, 4th ed. Editorial Hispano Europea, S.A. Barcelona, España.)

To construct the manufacturing process some authors (Tompkins et al., 2010) suggest beginning with the completed product and tracing the product disassembly to its basic components (Figure 6.7 and Figure 6.8).

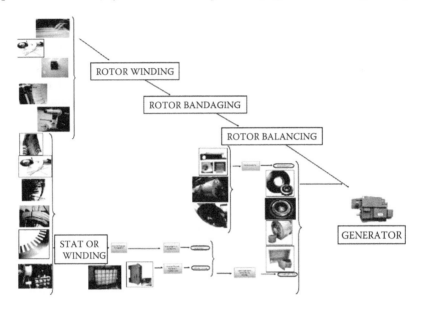

Figure 6.7 Assembly process for a wind generator. (From Indar Ingeteam, Bilbao, Spain. With permission.)

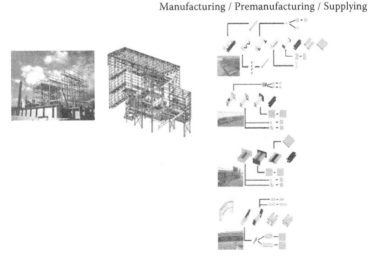

Figure 6.8 Assembly of a steel structure for a facility. (From URSSA. Vitoria, Brazil. With permission.) (Errasti et al., 2009.)

This production process "flow" shows the characteristics of a single, discrete product production, but the quantity of this product and other product families to be produced should be analyzed. Thus, **in designing the *flow*, the degree of commonality and modularity of product families should be considered.**

For this purpose, using group technology is appropriate, thus grouping parts into families and then making design decisions based on family characteristics. Groupings are typically based on part shapes, part sizes, material types, and processing requirements. This methodology seeks to aggregate operation volumes in order to select standardized processes and sort out appropriate machines, level of automation, and layouts.

The volume variety information that usually comes from the demand analysis in the business plan is very important in determining the layout type and machinery investment used for the flow design. For this purpose, the adapted matrix (see Chapter 6) that highlights layout alternatives depending on the product and process characteristics could be useful.

Flow Design Hierarchical Levels

The flow system in a facility is not just the flow of materials into the manufacturing plant "dock to dock," but also the flow of information and the logistic functions that condition these activities (Figure 6.9).

The tools and techniques used to analyze flow organization are explained in Chapter 7.

Flow planning hierarchy (Tompkins et al., 2010):

- Effective flow in workstations (Chapter 6)
- Effective flow within area (Chapter 6)
- Effective flow between areas or layout (Chapter 6)

Flow in Workstations: Is Automation Just a Reason of Volume or Something Else?

Workstation Flow Design with Production Volume Perspective

Figure 6.10 shows that one of the dilemmas when designing the manufacturing process is determining the level of automation. It is known that the production volume determines economic feasibility, and that it has to do with balancing investment cost versus operating costs in an automated, semiautomated, or manual manufacturing area.

Operating assembly systems with different levels of automation depends on the production volume.

Thus, the process engineering evolution allows manufacturing with different alternatives for each type of process (Table 6.4).

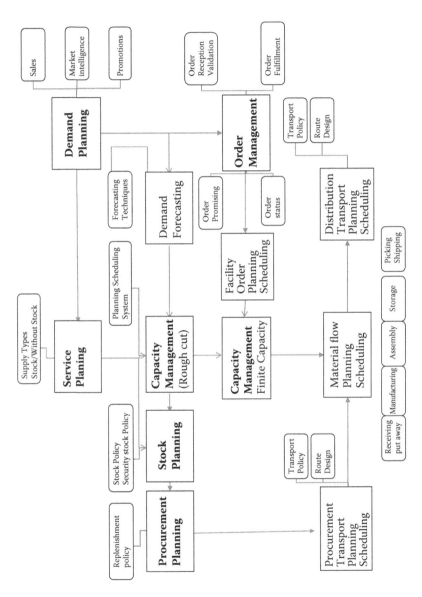

Figure 6.9. Logistic functions to be considered when designing a facility.

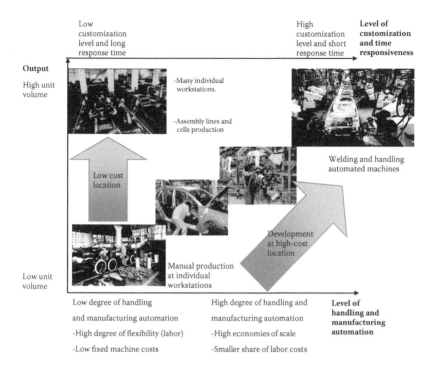

Figure 6.10 Operating systems with different levels of automation.

Workstation Flow Design Based on Quality

The use of alternative manufacturing technologies is also a key issue in the process engineering phase. Some authors (e.g., Corti et al., 2008) state that when transferring or redesigning a manufacturing process the automation level should be reviewed. The engineering department should address the reasons for this level of automation, especially in terms of quality and cost.

The level of automation should change if the main reason is "cost reduction in a high-wage country" when transferring the production line to a low-cost country. On the other hand, if the main reason is "quality improvement" and the reason for process automation has nothing to do with cost, the manufacturing process should have a similar design (Figure 6.11).

Similar to the manufacturing process, the logistic and warehousing process should be reviewed (Figure 6.12).

Table 6.4 Comparison of different welding system

| | Welding Robots | | |
	Automated Manufacturing and Handling	Manual Workstations Using Automated or Mechanized Handling	Manual Workstations with Partly Mechanized Handling
Investment volume	Very high	Medium	Low
Maximum products per day	3500 units	100 units	20 units
Maximum number of operations make for labor (polyvalence)	1–10 operations	10–50 operations	>50 operations

Source: Adapted from Abele, E., et al. (2008) Global production. A handbook for strategy and implementation. Springer, Heidelberg, Germany.

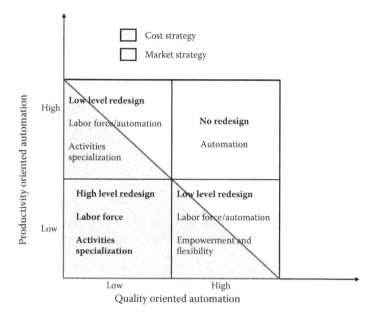

Figure 6.11 Process automation level redesign quality or cost oriented. (From Corti, D., et al. (2008) Challenges for offshored operations: Findings from a comparative multicase study analysis of Italian and Spanish companies. Paper presented at the 2008 EurOMA Congress, Groningen, The Netherlands. With permission.)

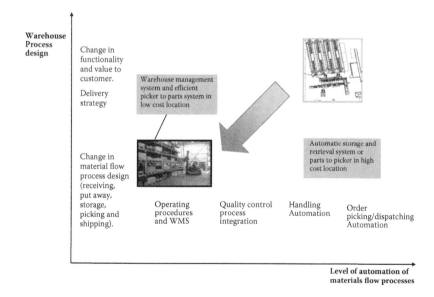

Figure 6.12 Example of alternative manufacturing methods with a change to the process design.

Workstation Flow Design Based on Variety and Volumes

Highly automated production can be attractive for large volumes and simpler methods when unit volumes are low and the variety is high (Table 6.5).

Volume eases the automation of the process, but the variety of products and the lack of communality could be a barrier. This paradigm highlighted the strategy of high-cost countries to reinforce high-value manufacturing in order to produce goods with a low level of standardization and highly customized designs (engineer to order). All of these paradigms are in the Manufacture 2020 Framework (Figure 6.13).

Flow within an Area: Production Areas

Areas are clusters of workstations that are to be grouped together during the facility layout process. This clustering could be carried out in two directions or axes:

- Workstations performing operations on similar products, or a **product layout**
- Workstations performing similar processes, or a **process layout**

Table 6.5 Concepts for manufacturing shoes called "alpargatas"

	Unique Shoes	Plain Shoes
Constraints	100 variants 500 units/year	10 variants 10 million units/year
Technology	Made by hand	Made by machine
Investment required	Low	High
Labor required	Many	Few
Production costs in:	High-cost country 30€	Low-cost country 3€

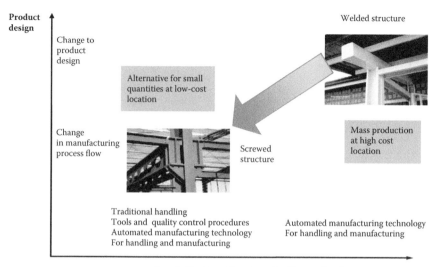

Figure 6.13 Example of alternative manufacturing methods with a change to the product design.

Product Layout versus Process Layout

As some authors have pointed out, a decision tree may be useful for deciding the process configuration (Cuatrecasas, 2009) (Figure 6.14).

When aggregating medium, volume-variety parts based on similar manufacturing operations, the machines required to manufacture are grouped together to form cellular manufacturing.

The clustering methodologies used to design cells (Tompkins et al., 2010) includes grouping parts together by listing parts and machines for the purpose of assigning machines to cells based on workload factors and general balance for a takt time or production rhythm. Then machines, employees, materials, tooling, and material handling and storage equipment are addressed in order to configure the layout of the cell.

A Company within a Company

Minicompany (Suzaki, 1993) or Fractal fabrik (Warnecke, 1996) is a name given to any unit of an autonomous and dynamic self-organized team within a company. Therefore, each minicompany has its customers, suppliers, bankers (bosses), and employees. These may include people within the company or from outside the company. This structure is referred to as a "holon," which is a combination of Greek words, *Holos* (whole) and *on*

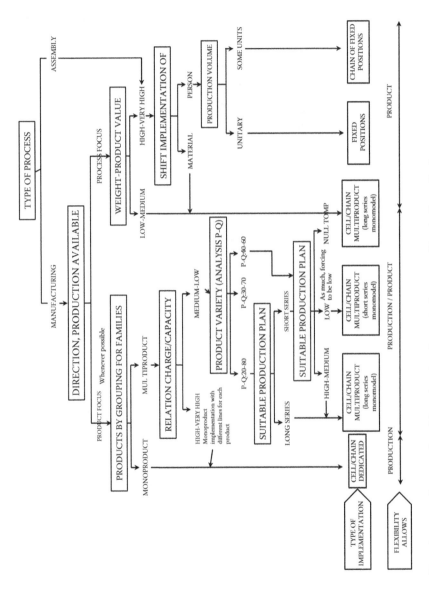

Figure 6.14 Classification of types of multiproduct implementation according to the criteria.

(individual). Here, a unit of an organization is a part of the whole and at the same time has an identity of its own with its subordinate parts.

The aim is for each of the units to show initiative and use its talents to the fullest while working. Also, because everyone is a customer and, at the same time, a supplier to others, practicing this idea brings benefit for all.

When somebody is the manager or the supervisor, for example, he or she is considered as the president of the minicompany. This person defines the mission and clarifies who works as the bankers, customers, suppliers, and employees. To accomplish the mission, moreover, this person sets objectives and develops business plans to accomplish them. Then, it will be communicated to key stakeholders so that the approach is in harmony with the others. Of course, the minicompany's progress should be regularly shared with them.

This management system, widespread in Japan and Europe, allows increasing competitiveness because the organization in the facility becomes flat and process-oriented, which facilitates the application of the principles of Continuous Improvement and Lean production.

Flow between Areas or Layout

Most facilities consist of a mixture of product and process areas that produce a large variety of products.

Layout Procedure

Muther (1981) developed a systematic procedure for layout planning. Based on the input data and an understanding of the roles and relationships between activities, a material flow analysis (from-to chart) and an activity relationships analysis (activity relationship chart) are performed. From these analyzes, a space relationship diagram is developed and, at this point, the layout alternatives are developed and evaluated (Figure 6.15).

Product/Volume Classification

Hayes and Wheelwright (1984) state that facility layout depends on product volume and variety and the characteristics of the production process. Cuatrecasas (2009) adds that, in addition to the classical job shop, the process shop and product shop of cellular manufacturing and flexible manufacturing systems also have to be considered (Figure 6.16).

Process Types

There are three main process types in a facility, which are classified by product and processes: A, V, and T types.

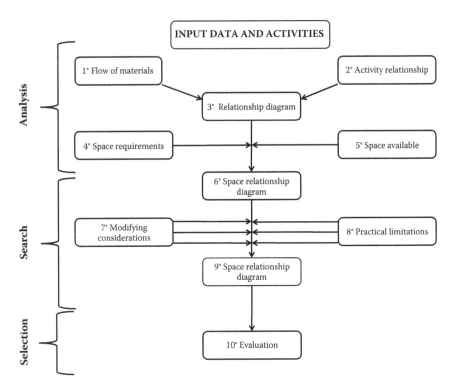

Figure 6.15 Muther's (1981) systematic layout planning (SLP) procedure.

In a "V" facility, there are few raw materials and the process is designed to obtain a greater quantity of final products, which differ in quality (different materials), but in application, it is quite similar. A steel factory is a typical V facility.

In an "A" facility, there are many different raw materials, components, and subassemblies that are turned into a small variety of products. An example of this type of facility is an airplane production facility.

In a "T" facility, the final product assembly has different combinations and configurations. In the first part of the process, the products are produced in a similar way (this first part could be a V or A shape), and in the second part, the assembly is carried out in many different ways. This facility concept is in tune with mass customization concepts where the product is basically customized in the last part of the process. Nevertheless, a company could have different product strategies in terms of customized solutions that impact the process variety and technology (Figure 6.17 and Figure 6.18).

In Table 6.6, the main characteristics of A, V, and T facilities are shown.

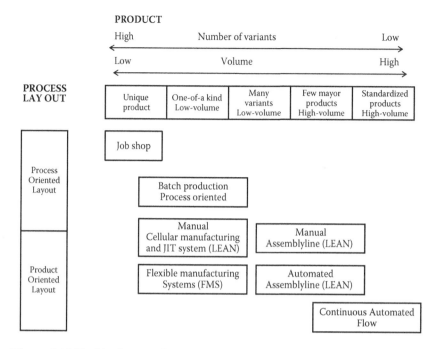

Figure 6.16 Facility layout alternatives depending on the product and process characteristics.

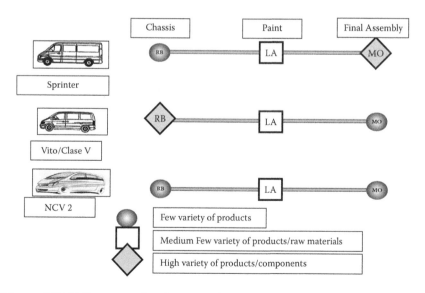

Figure 6.17 Different product/process alternatives depending on the product strategy. (Adapted from ACICAE (automotive component industries), Bizkaia, Spain, 2003.)

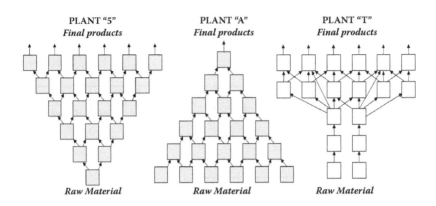

Figure 6.18 V, A, and T process types.

Table 6.6 Characteristics of A, V, and T facilities

V Facility	Very high investment in machinery and equipment
	High automation level
	Focused factory and product layout
	Flexibility limited and time response is in weeks
	Disadvantage: The efficiency of the facility is a key issue because they are usually cost-oriented facilities and have a floating bottleneck problem depending on the product mix and lot size
A Facility	High investment in machinery and equipment
	Different automation levels depending of the location, volume, and variety
	Primarily product or market focused factories and process mixed layout
	High responsiveness in time, quantity, and variety
	Disadvantage: Materials management is very complex and needs to be fixed
T Facility	High–medium investment in machinery and equipment
	Different automation levels depending on the location, volume, and variety
	Main process is assembly
	Primarily product or market focused factory and process mixed layout
	High responsiveness in a short time, quantity, and high variety
	Disadvantage: Materials management is very complex and needs to be fixed

Production Facility Design

The following characteristics have to be considered in the design process of a production facility (the production management systems will be detailed in Chapter 7):

- Technological level of the facility and the automation level of the processes
- Equipment and machinery distribution and location
- Cellular manufacturing design and balancing (flow concept)
- Design to demand capacity (work and mix production leveling, production and logistic capacity)
- Flexibility ("0" setup time), availability ("0" breakdowns), and quality ("0" defects)
- Autonomation (Jidoka machine/people dependency paradigm)
- Advance planning and scheduling system
- Visual management
- Production personnel organization
- Areas and workstation design
- Shop floor management design (mission, personnel, area, customer supplier relationships, processes, worksheets, communication, continuous improvement system)
- Key performance indicators (quality, cost, delivery, security, and moral) and improvement activities

In the past decade, under the "agile manufacturing"[*] concept paradigm and the reconfigurable manufacturing system, there has been a tendency to take into account capacity flexibility, which is the ability to vary the production level and transfer the production capacity from one product to another quickly and at a low cost. There are four ways in which this could be done, either alternatively or complementarily.

1. Flexible facilities, with "mobile" equipment and structures that can be disassembled, and reconfigurable product and services. This facility is appropriate under dynamic value chain conditions (see Chapter 5).
2. Flexible processes, with flexible manufacturing system, based on low investment equipment and high flexibility in product variety and customization, based on modularity and commonality.

---

[*] The term "agile manufacturing" is applied to an organization that has created the processes, tools, and training to enable it to respond quickly to customer needs and market changes while still controlling costs and quality.

3. Flexible workers, or workers who are skilled and polyvalent and have multiple capabilities to undertake different types of work and improvement (see Chapter 7).
4. Extended capacity, settling agreements with suppliers and even competitors, to outsource and subcontract part of the activity when needed.

References

Abele, E., Meyer, T., Näher, U., Strube, G., and Sykes, R. (2008) *Global production: A handbook for strategy and implementation.* Springer, Heidelberg, Germany.

Blackstone, J., and Cox, F. (2004) *APICS dictionary*, 11th ed. CFPIM, CIRM, Alexandria, VA.

Boyer, R., and Freyssenet, M. (2003) *Los modelos productivos.* Editorial fundamentos, Madrid, Spain.

Christopher, M. (2005) *Logistics and supply chain management: Creating value added network*, 3rd ed. Pearson Prentice Hall, Cambridge, U.K.

Corti, D., Egaña, M. M., Errasti, A., (2008) Challenges for off-shored operations: Findings from a comparative multi-case study analysis of Italian and Spanish companies. Paper presented at the 2008 EurOMA Congress, Groningen, The Netherlands.

Cuatrecasas, L. (2009) *Diseño avanzado de procesos y plantas de producción flexible.* Profit Editorial, Barcelona.

Errasti, A. (2006) KATAIA Diagnóstico y despliegue de la estrategia logística en Pymes. PhD diss. University of Navarra, Tecnun, San Sebastian, Spain.

Errasti, A. (2009) *Internacionalización de Operaciones.* Cluster de Transporte y logística de Euskadi, Diciembre.

Errasti, A. (2010) *Logística de almacenaje: Diseño y gestión de almacenes y plataformas logísticas world class warehousing.* University of Navarra, Tecnun, San Sebastian, Spain.

Errasti, A., Beach, R., Odouza, C., and Apaolaza, U. (2009) Close coupling value chain functions to improve subcontractors' performance. *International Journal of Project Management.* 27 (3): 261–269.

Errasti, A., and Bilbao, A. (2007) *Proyecto OPP Optimización Preparación de Pedidos*, Cluster de Transporte y Logística de Euskadi.

Flegel, H. (2004) Manufacture, a vision for 2020. Assuring the future of manufacturing in Europe. Report of the High-Level Group. November. Luxembourg. Official Publications of the European Communities.

Hayes, R. H., and Wheelwright, S. C. (1984) *Restoring our competitive edge: Competing through manufacturing*, 3rd ed. John Wiley & Sons, New York.

Hirano, H. (1998) *5 pilares de la fábrica visual: La fuente para la implantacion de las 5s.* Productivity Press, Cambridge, MA.

Muther, R. (1981) *Distribución en planta*, 4th ed. Editorial Hispano Europea, Barcelona, Spain.

Suzaki, K. (1993) *New Shop Floor Management: Empowering people for continuous improvement.* New York: Free Press.

Suzaki, K. (2003) *La nueva gestión del fábrica.* TGP Hoshin, Madrid.

Takahashi, Y., and Takashi, O. (1990) *TPM: Total productive maintenance.* Asian Productivity Organization, Toyko.

Tompkins, J. A., White, J. A., Bozer, Y., and Tanchoco, J. M. A. (2010) *Facilities planning*, 4th ed., John Wiley & Sons.
Warnecke, H. J. (1996) *Die fraktale fabrik*. Rowohlt Verlag, Reinbek, Germany.
Womack, J. P., and Jones, D. T. (2003) *Lean thinking*. Simon and Schuster, New York.

chapter 7

Planning/Scheduling Integrated System and Shop Floor Management

Jose Alberto Eguren, Carmen Jaca, Sandra Martínez, Raul Poler, and Javier Santos

> *What really matters is not arriving at the summit, but knowing how to stay there.*

Contents

Introduction

In this chapter, we discuss:

- Product attributes and applicable management principles
- Shop floor management basic and advanced level
- Continuous improvement organization

Product Attributes and Applicable Management Principles

Lean and Agile Manufacturing Principles

Over the past 20 years, small and large businesses have intensively applied the Lean approach. The origins of Lean manufacturing can be traced to the Toyota Production System (TPS). The core of this approach is the elimination of waste. Seven types of waste have been identified: overproduction, waiting, transportation and useless handling, bad or poor processing, excess inventory, useless motion, and the production of defective parts (Ohno, 1988). In order to eliminate the existing waste during the manufacture flow of a product or family of products, the value-creating steps must be distinguished from those not creating value. Lean manufacturing has evolved into Lean thinking principles whose implementation methodology can be summarized as: specify value by specific product, identify value stream for each product, make value flow without interruptions, let the customer pull value from the producer, and pursue perfection (Womack and Jones, 2005). These principles also are applicable to the manufacturing flow (Lyonnet, Pralus, and Pillet, 2010) of small and medium enterprises (SMEs) that seek to eliminate waste while ensuring continuous product flow. In addition to using value stream mapping (VSM) in order to allow visualization of the steps that create value, some authors propose the following mapping tools: process activity mapping, supply chain response matrix, production variety tunnel, quality filter mapping, demand amplification mapping, decision point analysis, and physical structure (volume–value) (Jones, Hines, and Rich, 1997) (Figure 7.1).

Christopher and Towill (2000) define **Agile manufacturing** as "the ability of an organization to respond rapidly to changes in demand, both in terms of volume and variety, to serve volatile and predictable markets." Harrison and van Hoek (2005) argue that Lean works best in stable, predictable markets where variety is low and there is a tendency to restrict Lean to factories.

Mason-Jones, Naylor, and Towill (2000) point out the product and demand attributes that make the implementation of Lean or Agile concepts more suitable (Table 7.1).

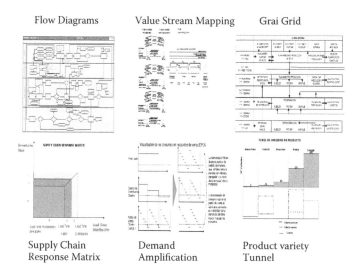

Flow Diagrams   Value Stream Mapping   Grai Grid

Supply Chain   Demand   Product variety
Response Matrix   Amplification   Tunnel

Figure 7.1 Complementary Lean logistics tools. (Adapted from Jones, D. T., et al. (1997) Lean logistics. *International Journal of Physical Distribution and Logistics Management*, 27 (3/4): 153–173.)

Table 7.1 Product and demand attributes and Lean or Agile concept implementation

	Lean	Agile
Typical products	Commodities	Fashion goods
Market place demand	Predictable	Volatile
Product variety	Low	High
Product life cycle	Long	Short
Information enrichment	Highly desirable	Obligatory
Forecasting mechanism	Algorithmic	Consultative

In contrast, some authors propose that both principles are applicable and state that the supply strategy and the order decoupling point determine where Lean is to be applied and where Agile is in the supply chain (see Chapter 3) (Figure 7.2).

Shop Floor Management Ramp Up Strategy: A Proposal for a Facility Improvement Roadmap

The authors of this book state that in the start-up process of a new facility the basic organization level should be reached, and that it is more crucial to implement the facility, equipment, and resources with less cost and less time deviation (see Chapter 9) than to implement the whole set of

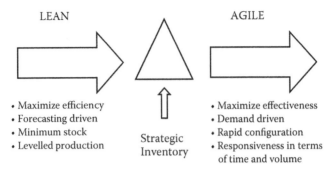

Figure 7.2 Application of Lean before the decoupling point and of Agile after the decoupling point.

tools and techniques related to the Lean approach. This approach, which does not consider a 100% system implementation at one time, has been also considered by other researchers when implementing Lean management in existing semi-automated manufacturing lines (Bowler et al., 2010; Martinez et al., 2012)).

Some authors (Taylor and Brunt, 2010) have already identified the Lean tools sequence while implementing Lean in a facility. This ordering of the applicable tools and methodologies is an interesting contribution.

Nevertheless, this approach does not take into account that these tools, in fact, are methodologies that have cultural implications and embedded values (Nakano, 2010), and that if they are not implemented gradually it could be unsustainable. Thus, training and the commitment of top and midmanagers are prerequisites to changing people's mentality and leading the change in management.

Prajogo and Sohal (2004) define sustainability "as the ability of an organization to adapt to change in the business environment, to capture contemporary best practice methods. and to achieve and maintain superior competitive performance." There are different factors that may inhibit or enable sustainability (Jaca et al., 2010), such as:

1. Managerial commitment
2. Key performance indicators
3. Program objectives linked to strategic goals
4. Achievement and implementation of results
5. Appropriate methodology
6. Specific program resources
7. Involvement of task force
8. Adequate training
9. Communication of the results
10. Increase in employee involvement

Table 7.2 Application of Lean management tools

Start-Up	1st S, 2nd S, workplace basic organization				
	Corrective maintenance				
	Planned maintenance in bottlenecks				
	Quality control procedures				
	Team creation				
Team and Equipment Stability	Standard operations sheets				
	Micro layout design				
	Planned maintenance program				
	Autonomous maintenance				
	Overall equipment; efficiency measurement and key problems elimination				
	Problem solving; techniques				
Improvement	Standardized manufacturing visual procedures				
	4th S, 5th S				
	Total quality control advance techniques (Poka Yoke, Jidoka)				
	APS planning/scheduling system (pull/push, Kanbans, Andon)				
	Continuous improvement programs Kaizen (layout change, waste elimination, setup reduction)				
	Continuous improvement programs with key suppliers				
	Benchmarking programs				
Excellence	Fractal companies management				
	Reengineering projects				
	Key suppliers integration				
	Factory role definition and upgrading				
	Key customers integration				

Source: Adapted from Taylor and Masaaki.

11. Promoting of teamwork
12. Facilitator to support the program
13. Appropriate areas for improvement
14. Adaptation to the environment
15. Participant recognition

Therefore, it is necessary on the one hand to adapt traditional Lean tools and practices in order to address current ramp-up process limitations (see Chapter 9) and, on the other hand, propose a three-stage Lean implementation strategy that is based on the Toyota Production System. The first two stages are more oriented toward gaining operational stability and equipment availability and the third step is more oriented toward Build-in quality and Just-in-Time (JIT) principles, while considering the human factor in all the processes (Figure 7.3).

This strategy enables maintainable quasi-Lean processes in the short term, permitting a Lean corporate culture to develop, which in turn leads to long-term sustainable Lean processes.

Keeping in mind these steps, which are based on the Toyota Production System Lean tools and also on the Continuous Improvement roadmap

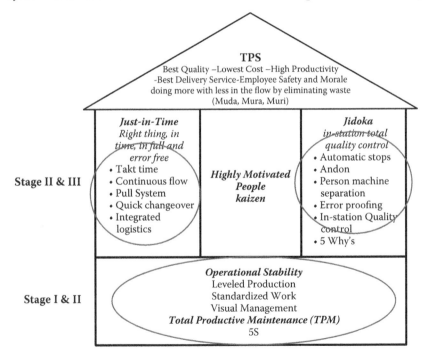

Figure 7.3 Proposed three-stage Lean implementation strategy taken from the Toyota Production System (TPS).

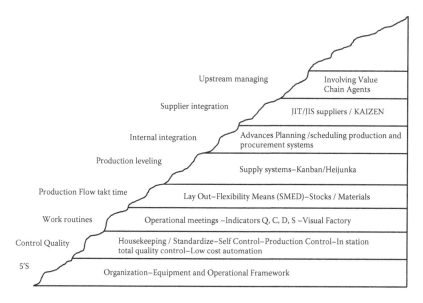

Figure 7.4 Continuous improvement roadmap. (Adapted from Masaaki (1990).)

proposed by Masaaki (1990) (Figure 7.4), a new roadmap is exposed in the following text.

This roadmap, which has been developed by the editor of the book and the contributors of this chapter, proposes four stages (basic, stability, improvement, and excellence) in order to manage the ramp-up strategy in an effective way.

Stage I: Start-up basic organization

- Manufacturing process procedure analysis and mapping (Ishiwata, 1997)
- Micro layout design:
  - First S Seiri-Sort: in production and logistic areas ensuring each piece of equipment and the handling system, tools, and work-sheets in a workplace are in their proper place or identified as unnecessary and removed.
  - Second S Seiton: Arrange materials and equipment so that they are easy to find and use.
- Basic equipment availability:
  - Safety in workplace and corrective management oriented toward process analysis and mapping.
  - Planned maintenance analysis and mapping for bottlenecks and critical equipment.

- In-station quality control procedures in key product functionalities.
- Creation of Operations Improvement team and propose tools in order to test the commitment of managers and engineers.

Stage II: Team and equipment **stability**

- Manufacturing process procedures put in the workstations and areas in a visual way.
- Micro layout design:
  - Third S Seiso-shine: Repair, clean, and shine work area, cleaning inspection checklists and schedules.
- Equipment availability:
  - Safety and ergonomic control and improvement activities.
  - Conditional preventive maintenance implementation.
  - Total productive maintenance (TPM) level four (corrective maintenance and preventive maintenance/self-maintenance).
  - Start up with autonomous maintenance.
- Quality control procedures based on person–machine separation.
- Quality and service key performance indicators (KPI) in production areas and quality circle teams with employees oriented toward security, quality, and service.
- Advanced planning and scheduling (APS) Planning/Scheduling system procedure based on constraints and supplier bottlenecks (service first, then stock and utilization improvement).
- Productivity overall equipment effectiveness (OEE) KPI indicators and supplier quality/service assessment monitored by managers.

Stage III: Start-up Improvement

- Standardized manufacturing visual process procedures.
- Micro layout design:
  - Fourth S Seiketsu-Standardize: Formalize procedures and practices to create consistency and ensure all steps are performed correctly. Everyone knows what they are responsible for doing, and when and how. See production status at a glance.
  - Fifth S Shitsuke-Sustain: Create awareness of improvements, management support for maintenance, training, and rewards.
- Equipment availability:
  - Simplify equipment inspection (visual).
  - Standardize corrective and preventive maintenance procedures.
  - Measure and improve OEE programs related to maintenance management.
- APS Planning/Scheduling system improvement:
  - Order decoupling point pull–push implementation.

- Visual and leveled replenishment systems (kanbans, capacity requirements planning [CRP]).
  - Visual production control systems (Andon).
- Quality, Productivity (OEE), and Service KPI indicators in production areas and teams oriented toward competitive improvement of operations.
- Supplier development to improve quality and service.
- Benchmarking with lead factories and Industrial Reference Productions Systems maturity model assessment and Improvement programs.

Stage IV: Excellence

- Improve and revise the standardized manufacturing visual process procedures.
- Micro layout design:
  - 5S Shitsuke-Sustain: Create awareness of improvements, management support for maintenance, training, rewards.
- Equipment availability:
  - Early management of new equipment and engineering maintenance development.
  - Reinforce autonomous maintenance.
  - Develop predictive maintenance in key equipment.
  - OEE benchmark with lead factories and maintenance cost objectives deployment.
  - Preventive maintenance reinforcement and quality maintenance development.
- APS Planning/Scheduling system improvement:
  - Integration of the APS with the ERP and Production control system.
  - Integration of the APS with customers.
  - Integration of the APS with key suppliers.
  - Visual and leveled replenishment systems (kanbans, CRP).
- Key supplier integration in quality, service, and cost and new supplier development.
- Definition of factory role within the manufacturing network.
- Key customers, key account managers, and also facility operations strategic alignment with customers.
- Plant/facility reconfigurations to new customer needs (volume, flexibility, variety, etc.).
- Best practices standardization and communication to the group.

However, all of these stages must be adapted taking into consideration the characteristics and features of the organization. Additionally, Jaca (2011) proposes a framework model to implement all of the tools mentioned above through improvement teams (Figure 7.5).

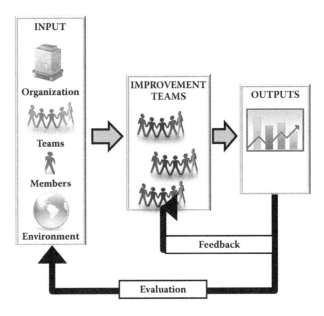

Figure 7.5 Model for continuous improvement program.

The model incorporates the use of improvement teams as a way of making improvement sustainable. In this model, the environmental context (social, political, and cultural) is an important input for the continuous improvement program.

Applicability and Transfer to Offshore Facilities Shop Floor Management

Some authors have stated (e.g., Olhager, 2003) that an operations strategy should be based on a strong systematic and standardized way of working combined with empowered shop floor teams that drive continuous improvement in that standardized work.

Lean, TPM, continuous improvement, and supply chain management are key methodologies (some authors would say management principles) to improve the capability of a facility to improve overall performance and empower shop floor teams. Nevertheless, shop floor management implementation also needs a ramp up process where the management tools that guarantee the availability of equipment, the quality of the product process, the efficiency of the facility, and supplier and delivery conditions are put in place gradually. This process should be piloted by the operations strategy managers, and for this purpose some authors have already proposed how to focus this kind of project from the knowledge management point of view (Baranek, Hua Tan, and Debnar, 2010).

Therefore, shop floor management implementation looks like a key problem, especially taking into account that when transferring manufacturing to more appropriate locations, only a fraction of managers were satisfied with the way knowledge and experience (Nonaka, 1994) were transferred within their organizations (Ruggles, 1998; Szulanski, 2003; Errasti and Egaña, 2008). The process of transferring knowledge and best practices proposed by Szulanski (1996) includes initiation, implementation ramp up, and integration with four milestones: definition of transfer speed, decision to transfer, first day of use, and achievement of satisfactory performance.

Baranek, Hua Tan, and Debnar (2010) also point out that the capacity to absorb new production practices efficiently depends on the environment of the receiving site. An existing manufacturing environment (Brownfield) challenges the current paradigms and, hence, makes the transfer of management principles more difficult, while new manufacturing facilities (Greenfield) have no preconceived ways of thinking and doing, so they are more capable of absorbing knowledge and practices.

Flow Organization

Production Planning and Scheduling Systems

Once the availability, quality, and flexibility problems are solved, the main origins of waste in the production process are under control. Nakajima (1998) identified and classified the main losses related to availability, performance, and quality. He established the "six big losses": (1) poor productivity and lost yield due to poor quality, (2) setup and adjustment for product mix change, (3) production losses when temporary malfunctions occur, (4) differences in equipment design speed and actual operating speed, (5) defects caused by malfunctioning equipment, and (6) start-up and yield losses at the early stage of production.

Following the Masaaki roadmap for Continuous Improvement, once the Total Quality Control, the availability of machines, and the system's flexibility to produce in lot size are reasonable, **the system flow has to be considered**.

Crucial in the application of Lean or Agile principles to flow organization is the step-by-step method of production planning (Haan et al., 2010): In much of the hierarchical planning to forecast systems, aggregated planning is refined to determine the capacity for the next short period (1–3 months), and from then on planning is to order in a shorter fixed period or master planning scheduling (2–4 weeks) for customized products and continues to be even more refined forecasts for commodities, which are finally to order with narrower time windows (24–72 hours). These refined planning steps more or less equal customer orders because of their accuracy and these orders can be produced on time in the 100% controlled production processes (Figure 7.6).

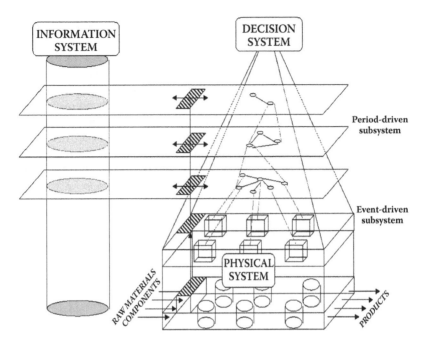

Figure 7.6 Hierarchy of the operational planning decision systems that condition the physical system.

A not very well understood methodology to accomplish this complexity is the GRAI method (Doumeingts, 1984) and the GRAI grid (Errasti et al., 2006, 2012). Among the different instruments used to analyze the decision system of an enterprise, the GRAI method provides a classification of enterprise decisions based on the two dimensions: decision level and functions. Enterprise decisions are represented as decision centers located in specific positions on the GRAI grid. The decision activities are mapped on the GRAI nets, which show the decision flows.

To manage an enterprise, many decision centers operate concurrently. To coordinate and synchronize decision making, decision flows and feedback connect decision centers together. Operational business processes, such as production and procurement planning and scheduling are driven by a decision-making function hierarchy. The different decision-making functions consider different sets of resource issues (time, capacity, etc.), different time horizons periods, and different decision centers (production, purchasing, etc.). This technique enables the planning/scheduling decisional level of the factory (Figure 7.7 and Figure 7.8) and the extended supply chain, including suppliers, to be monitored.

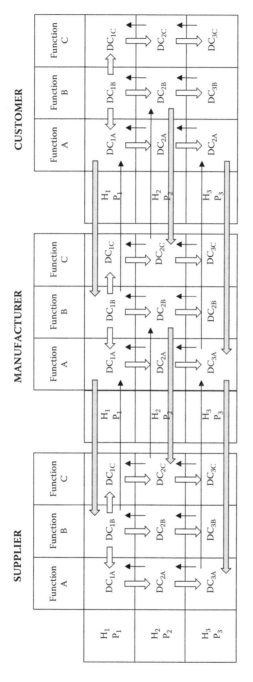

Figure 7.7 Hierarchy of extended planning decision systems including suppliers and customers.

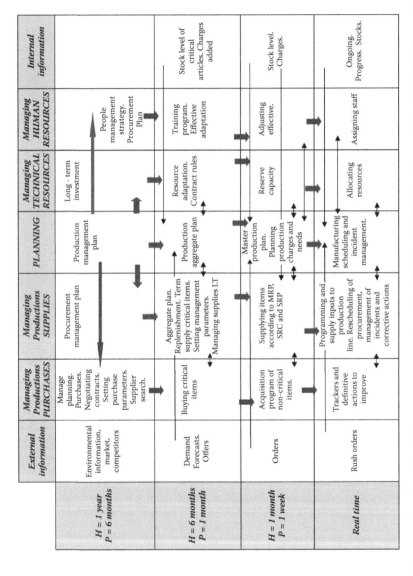

Figure 7.8 Factory planning/scheduling systems monitored in GRAI grid. (Adapted from Poler, R. (1998) Dynamic Analysis of Enterprise Decisional System in the frame of the GRAI method. Doctoral Thesis. Polytechnic University of Valencia. With permission.)

Advanced Planning and Scheduling Functionalities

There are many different kinds of decisions to be made in an enterprise or network. It is usual to classify decisions in three levels: strategic, tactical, and operational. The criteria for classifying a decision into one of these levels can be temporal coverage and logical granularity. In terms of temporal coverage, strategic decisions cover a longer time horizon than tactical decisions, and it is the same respect for the operational decisions. Concerning logical granularity, the level of detail of data and variables becomes increasingly finer when decisions are more operational than strategic.

Even though some authors have identified the great advantage that manufacturing companies have when implementing automatic real-time planning and control systems (ERP-APS-MES) (Arica and Powel, 2010), this book considers production planning and control (PPC) systems at the individual plant level and reflects on their integration within the supply chain.

For this purpose, ERP (enterprise resource planning) systems have to be adapted to allow advanced planning and scheduling (APS) systems to be implemented. These systems aid in the creation of a support mechanism for planning and decision making at the operational planning level.

These systems are an evolution that integrates:

- **MRP** (materials requirements planning): A material planning method that uses a production bill of materials, inventory management techniques, and master production scheduling to determine material needs and replenishment timing.
- **CRP** (capacity requirements planning): Finite capacity resources, such as manpower, machine hours, and procedures.
- **DBR** (drum–buffer-rope): Based on systemic principles from OPT (optimized production technology) and the theory of constraints applied to operational planning or DBR.

These systems have the following functionalities:

- **Handle the capacity constraints of different resources,** such as materials, labor, transport, and plants. Thus, APS systems when combined with ERP systems replace the infinite capacity logic of MRP and allow for explicit, capacity-constrained production planning and control. They do not only consider just shop floor capacity constraints, but also inventory and bill of material constraints, inventory stocking and replenishment levels, and order generation policies (Turbide, 1998).
- **Schedule in the constraint** by giving a first rough cut capacity and a detailed capacity analysis in the bottleneck, subordinating and scheduling outwards from a bottleneck.

- **Allow availability** to promise that the system is able to allocate supply to customers.
- **Planning multiple manufacturing plants** by defining the timing and volume of purchases of longer lead time raw materials that affect planning decisions (signal to purchase).
- **Planning multiple manufacturing plants.** If the business has a limited number of manufacturing sites, this function may be combined with tactical business-wide planning. Retain the original due date and calculate the renewed delivery date, each time scheduling is performed, resequencing activities and orders.
- **Allow production level and smoothing policies** that play an important role in the JIT system.
- **Allow shifts and workstations to be configured and designed** in order to have different production levels that are adaptable to different demand levels.

Planning and Scheduling System as an Integrator of the Extended Enterprise

Another important difference in how customers value products is *customization*. Creating unique products is only possible if the customer can influence the properties of the products, meaning that the product, to some extent, is engineered to order. In this context, some parts of the logistic activities are performed as the customer is waiting, but it is also the case that some preceding activities may have to be performed on speculation due to the fact that the production lead time is longer than the required delivery time (Wilkner and Rudberg, 2005). A concept frequently used to capture this aspect of operations strategy is the customer **order decoupling point** (CODP), which decouples operations in two parts (Hoekstra and Romme, 1992). Upstream of the CODP, activities are performed to forecast, which is known as push production, and downstream they are performed to customer order, which is known as pull production. Typically, the four CODPs are defined as: engineer to order, make to order, assemble to order, and make to stock (Olhager, 2003).

As an example, we can see different decoupling points and alternatives in a distribution network in Figure 7.9.

As another example, the decoupling point enters the facility, which generally is the same as **order penetration point** (where materials are assigned to customer orders), determines the pull–push logic, and the locations and management of stock buffers (Figure 7.10).

This way of managing production allows for the leveling and smoothing of the production upstream from the supply chain, even the suppliers. In Figure 7.11, we see an assembly to order system with a buffer that

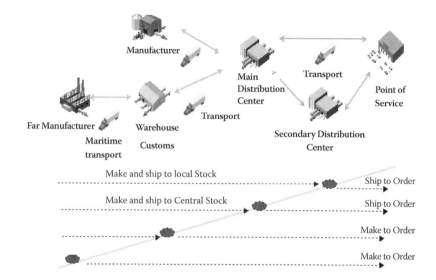

Figure 7.9 Different decoupling point alternatives in a distribution network.

cushions the variability of customer demand and allows leveling demand and orders to long lead time suppliers.

The planning/scheduling production and procurement system also can be applied to improve quality service and reduce integral cost in multiproduct assembly lines.

This system consists of applying the "setup wheels" concept. This allows great consumption articles to be rotated as much as possible by taking into account the bottleneck capacity, the master planning and scheduling period and horizon, and the delivery strategy (make to stock or make to order). These types of production systems are typical in the automotive and household appliance supply chains (Figure 7.12).

Advanced planning and scheduling systems can be applied when they are integrated in the planning system by simulating different batching or grouping of orders depending on the commonality of materials and setup times and showing the performance in terms of stock and setup times (Figure 7.13).

Moreover, advanced planning and scheduling systems **allow a production planner to decide the right production level among different configured shifts and workstations to have different production levels adaptable to different demand levels. In Figure 7.14, we see the different alternatives of shifts and labor balancing in four assembly cells.**

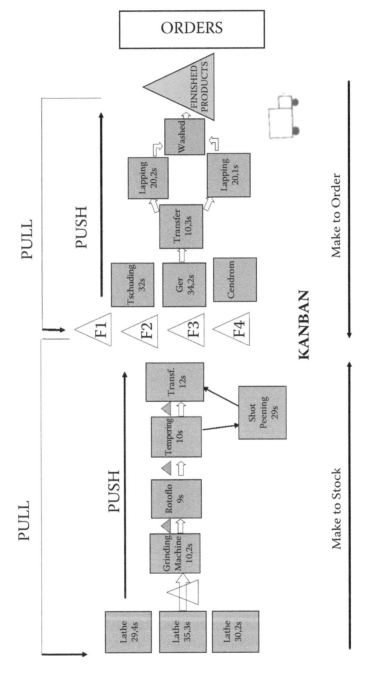

Figure 7.10 Example of a two-tier facility and the decoupling and order point.

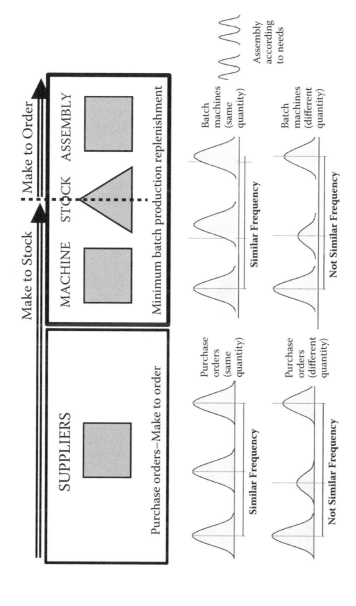

Figure 7.11 Assembly to order system and suppliers leveling and smoothing.

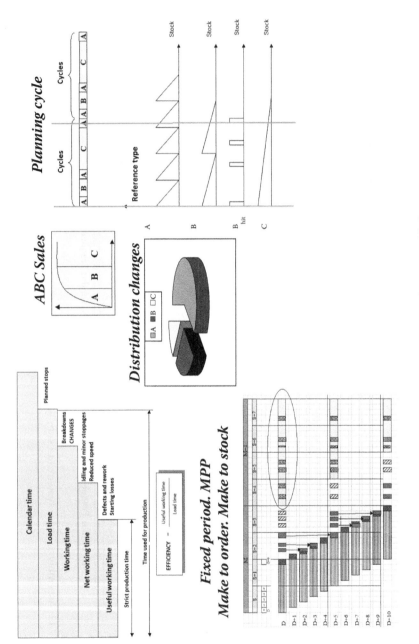

Figure 7.12 Example of a setup wheels planning system in a multiproduct assembly line.

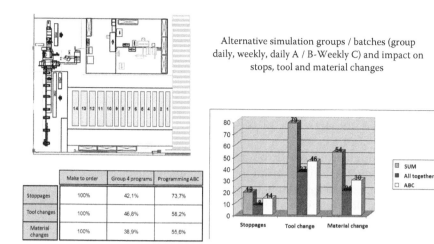

Alternative simulation groups / batches (group daily, weekly, daily A / B-Weekly C) and impact on stops, tool and material changes

	Make to order	Group 4 programs	Programming ABC
Stoppages	100%	42,1%	73,7%
Tool changes	100%	46,8%	58,2%
Material changes	100%	38,9%	55,6%

Figure 7.13 Simulation of different batching and grouping alternatives in a manufacturing cell.

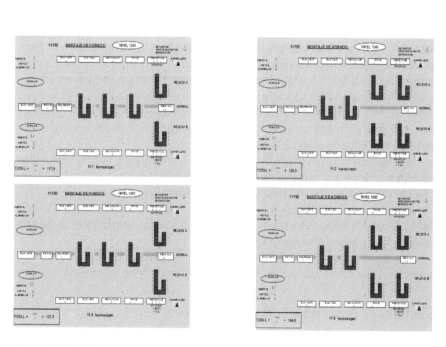

Figure 7.14 Different shifts and personnel balancing alternatives in four assembly cells.

Continuous Improvement Organization

Continuous Improvement (CI) (Masaaki, 1986) is defined as the process of steadily and gradually improving the different areas of a company, seeking greater productivity and competitiveness.

The objectives of CI can be summarized as:

1. To focus the activities of the company in improving the performance of processes
2. To gradually improve through incremental innovation
3. To conduct activities through the involvement of everyone in the company, from top management to production workers
4. To promote creativity and learning, developing an environment that promotes personal promotion

Descriptions of the successful implementation of CI programs have been largely reported, as well as the implementation of different models to achieve CI processes in companies (Jorgensen, Boer, and Gertsen, 2003; Bateman and Rich, 2003; Bateman and Arthur, 2002; Bessant, Caffyn, and Gallagher, 2001; Upton, 1996). However, a number of authors have expressed difficulties in sustaining CI over the long term, especially after an initial period of two or three years (Bessant and Caffyn, 1997; Schroeder and Robinson, 1991). It has been documented that the results, in terms of routines and acceptance of the system by the organization, happen after a period of five years (Jaca et al., 2010). Continuous Improvement programs have been traditionally used by most mature industries. These industries have to face CI to increase their production efficiency. However, it was found that there is still room for improvement to increase production efficiency and process sustainability so that new models need to be developed (Eguren et al., 2010; Jaca, 2011).

In this direction, studies by Corti, Egaña, and Errasti (2008) argue that in order to increase productivity, the strategy of organizational CI may change depending on the location of its production plants.

Eguren et al. (2010) have proposed a sustainable CI model called MMC-IKASHOBER, which could be appropriate to **implement once the facility has passed the ramp up process.**

For the design of this model, the basic elements of the CI and the key aspects of the OL (Organizational Learning) have been identified.

Continuous Improvement Basic Elements

There have been many studies that have identified the elements related to CI, which should be taken into account when designing a CIM (Continuous Improvement Model). As a rule, most studies agree on the

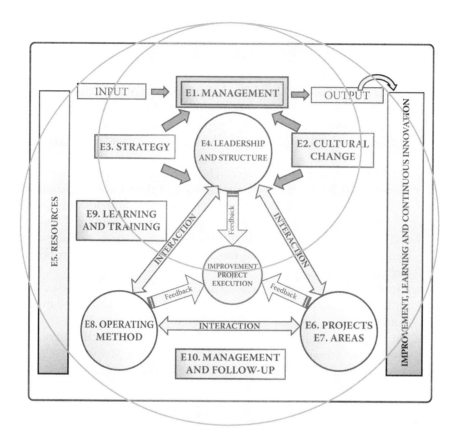

Figure 7.15 CI basic elements.

aspects mentioned, although each highlights the importance of different elements according to the approach of the study carried out. The studies mentioned mostly agree on the elements to be taken into account and, as seen in Figure 7.15, they include the following:

- **E1: Commitment of the management.** In order to address a CIP (Continuous Improvement Process) it is necessary to have management support and involvement (Deming, 1986; Juran, 1990; Feigenbaum, 1986), with a management style that encourages the CI in the organization routines and processes (Curry and Kadasah, 2002), and to this end it is necessary to define a lead team to direct the CI process (Crosby, 1979; Szeto and Tsang, 2005; Magnusson and Vinciguerra, 2008).
- **E2: Company culture.** In order to induce cultural change, it is necessary to develop new behavior and routines for all members of the organization that involves continuous learning and generating a

high level of organizational learning (Bessant, Caffyn, and Gallagher, 2001). Overcoming resistance to change using communication and ensuring that the CIP creates benefits (Deming, 1986; Juran, 1990; Feigenbaum, 1986; Kotter, 2007).

- **E3: Strategy.** The CIP should be a strategic part of the plan of operations which helps to create benefits on the different strategic levels through taking skills learned from the CI and turning them into routines (Hyland, Mellor, and Sloan, 2007; Bessant, Caffyn, and Gallagher, 2001).

- **E4: Leadership and structure.** There must be an organizational structure dedicated to CI that takes care of designing the strategic goals and responsibilities, managing the budgets, and designing and applying a system to measure the improvement (Middel, Gieskes, and Fisscher, 2005). A structural model that integrates seamlessly into the organization is what the Six Sigma methodology proposes (Schroeder et al., 2007).

- **E5: Resources.** Financial resources, release from other tasks for those taking part in the CIP, and time for training must all be made available (Szeto and Tsang, 2005; Bateman, 2005; Juran, 1990).

- **E6: Projects.** The projects must be clear, powerful, specific, feasible, realistic, and measurable, and must have a strong possibility of succeeding. They should be management selected, in line with strategy, and should help toward generating value for the client (Goh, 2002). The projects should be used as learning elements and the level of difficulty of the project and the skills to be developed, which are expected from the teams working, should be taken into account (Tort-Martorell, Grima, and Marco, 2008).

- **E7: Areas.** Focus must be placed on the critical processes, process improvement must be thoroughly carried through, and the impact of this on the general context of the organization must be taken into account (Garcia-Sabater and Martin-Garcia, 2009).

- **E8: Operational method.** It is necessary to have in place an RSP operational method based on the PDCA (plan-do-check-act) cycle and its respective tools (Pyzdek, 2003; Middel, Gieskes, and Fisscher, 2005; Magnusson and Vinciguerra, 2008).

- **E9: Training.** Specific training must be designed based on the skills and behavior to be developed at all levels. The contents of this training should basically involve the operational method and its respective tools for improvement both for related techniques and the personal relationship between people, such as communication, problem solving techniques, and teamwork. (Hoerl, 2001). The learning process should be based on the Kolb cycle (Kolb, 1984) and the learning-by-doing educational model (Upton, Bowon, 1998).

- **E10: Management and follow-up.** A CIP follow-up process should be established (Bateman, 2005), defining the indicators based on the efficiency, effectiveness, and learning developed in the CIP (Wu and Chen, 2006).

Organizational Learning Aspects

In addition to this, the importance of developing capacity for learning has been highlighted in the definition of an OL model. The model is based on the fact that in the organizations adopting an MMC-IKASHOBER, two activities clearly requiring continuous learning can be identified (Pozueta, Eguren, and Elorza, 2011):

- Learning to solve problems
- Learning to avoid problems

In order to learn to solve problems it is necessary to have problems in the hands of those responsible for learning this skill or ability. Each of these persons starts from a Mental Model (Senge, 2005) on how to perform this activity, a model which is, in general, neither very elaborate nor systematic and they will modify and perfect this as they gain experience (Kolb, 1984). As such, this involves skills building based on problem solving procedures.

In order to learn how to avoid problems, the starting point is a Mental Model of the system addressed (a corporate process, a machine, etc.) with agents and relationships, and this involves acquiring new skills regarding the way it works, which changes the decision-making process. As a consequence, existing operational procedures are modified.

In Figure 7.16, we can see the learning model developed for the present model and here we can see that learning is generated in two spheres. The first relates to the area itself where the improvements are applied, whereas the projects are implemented using the DMAIC (define, measure, analyze, improve, and control) methodology, and, as the goals are achieved, the new routines are identified which help to change behavior at the operational level for the area process to be more efficient and more effective.

The second sphere involves where the CIP learning takes place because, as the projects are implemented, the results of the application of the methodology, the development of the activities, and so on are assessed.

The development of the improvement process in itself is observed and new routines are identified that assist individual skills building and which lead to the CIP developing in a more effective and efficient manner. The process that follows for this is shown in Figure 7.17 and is the following:

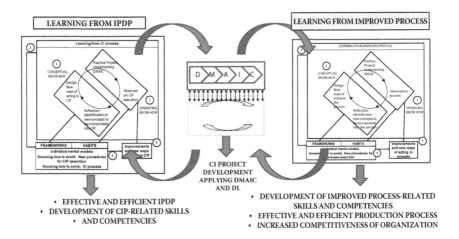

Figure 7.16 Organizational Learning (OL) Model.

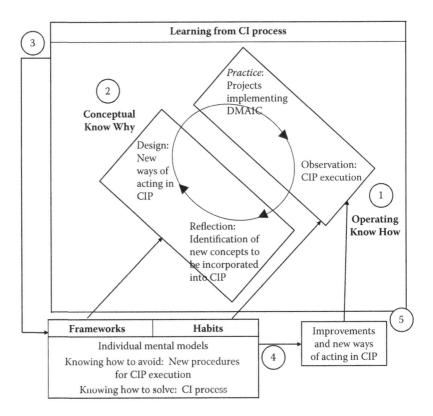

Figure 7.17 Learning cycle developed with the MMC-IKASHOBER.

- The individual applies the methodological skills acquired (DMAIC) with the aim to build up know-how or problem-solving skills. The continual application of methodological skills on different experiences to projects leads to variations and improvements in the way that the individual responds and manages to deploy the know-how to the MMC-IKASHOBER. For individuals to achieve the skill to carry out a task, it is necessary to design routines that through repetition lead to acquisition of the given skill. The given routines are shared and form part of the organizational knowledge base on MMC-IKASHOBER.

- In order to be able to design routines, it is necessary to know the know-why, i.e., why things work as they do. The individuals who know the CIP should design routines so that when the CI projects are implemented the teams try to build up problem-solving skills based on repeating these routines. Designing CIP routines is the trainer's responsibility.

- When in a project the root cause is identified and it is known "why" the process does not work, the know-why is understood. So the team, with this knowledge, has to make the affected environment change and to this end must design routines to facilitate the skill that enables the process to be improved.

- The persons who acquire the skill go on to a different level and a modification to the Mental Models is produced. If they understand why they do certain things, they can even improve their processes without the need for trainers as if it were an automatic process.

- The individual and their team, using their process knowledge, draw up or modify routines that are to be incorporated into the organizational knowledge base.

It must be taken into account that the MMC-IKASHOBER routines are drawn up by an expert team in PR methodology, which has a strong understanding of the know-why of MMC-IKASHOBER. This means it is necessary that it be a CIP trainer who improves the CIP and in each organization a start will be made in certain areas according to the goals.

The routines identified for the system affected are drawn up by the team working on the project that has a strong understanding of the know-why of the process on which the project is deployed.

A particular application of the previous model could be summarized as follows. The **conceptual aspects of the model** include:

- **Strategy:** From a strategic standpoint, the company must be prepared to apply this model, previously having taken into account the strategic culture of the company, the commitment of the management, and the involvement of the employees to adapt these aspects.

APPLYING THE MMC-IKASHOBER IN AUTOMOTIVE AUXILIARY AND HOUSEHOLD APPLIANCE COMPANIES BELONGING TO THE MONDRAGON COOPERATIVE GROUP

The aforementioned group is the largest business group in the Basque Autonomous Community and the seventh in Spain.

The model was implemented from 2007 to 2010 in 12 different companies in the automotive and household appliance sectors with different production plants and in which the CI is a strategic element and is deployed top-down.

First of all, the Management has to know the process and its role as well as the main features of the training programmes to be implemented. Secondly, the need to carry out a prior effort to choose the persons to lead the projects, recommending methodical persons who are persevering when faced with difficulties, with diagnostics skills based on continual questioning and statistical thinking and having communication and team leading skills as well as the time available, were highlighted. Thirdly, the operating routines and the skills which should be developed by the Teams in order to be able to deal with the projects effectively and efficiently were identified and applied. Then, the activities which took up most of the work were data gathering and analysis, diagnostics, and communication. Finally, routines were carried out which are oriented towards the use of a scientific approach, the statistical thought process and proof-based communication.

The most important conclusions achieved were:

- The statistical thinking is the most complicated to instill. The research team believes that promoting the capacity to imagine evidence is a good path to achieve the end as this this enables the decision-making cycle to start up in a planned fashion.
- The activity of developing a robust standardization system took up a lot of work. The goal of this task was systematically integrate the improvements identified.
- Communicating the results to "convince" the Organization and bring about organizational training which leads to a behavioral change is an important activity.
- The communication must be based on showing evidence, founded on good data-gathering, which brings in new ways to manage the areas addressed.

- **Project types:** The projects faced must be unceasing, manageable, and oriented to improvement. The suitable ones are through teamwork dynamics, with an established method with a long-term and endless process–project approach.
- **Theoretical training:** When designing a training program, it is essential to make employees aware of their role in the CI, and increase their ability to analyze, measure, and improve processes. Some authors join the training needs of the individual and the organization to tackle a project of CI.
- **Training in action:** It is important to continuously train participants in new techniques in order to be able to solve more increasingly complex problems and to increase the interest of the participants in the program itself.

The operating procedure of the model includes the operational method, which is the method followed based on the Six Sigma (SS) DMAIC methodology. Six Sigma is defined as the process or methodology through which improvements are obtained in the Six Sigma strategy. Six Sigma is an organized and systematic method for strategic process improvement based on statistical techniques and scientific methodology.

In this research project, due to the findings of previous studies and to strengthen the implementation of the Six Sigma methodology, it has been considered opportune to amend the methodology addressed and to propose using the following seven phases:

- **Phase F1 (Identify the problem):** In this phase, you must ensure that the project impacts in strategic areas of the company.
- **Phase F2 (Collect and analyze data (starting point)):** Deep initial diagnosis is that of drawing on validated metrics and rigorously defined defect/opportunities.
- **Phase F3 (Analyze the causes):** Identification of the few X variables, the root causes of the problem, in a rigorous way, confirmed by experimental evidence.
- **Phase F4 (Plan and implement solutions):** Evaluation of the selected improvement ideas in a controlled and rigorous pilot testing experience, monitoring side effects, and other elements that help to modify the ideas.
- **Phase F5 (Test results):** Check that the improvement is sustained over time, measuring an indicator agreed with all the parts concerned.
- **Phase F6 (Standardize results):** Validation of the operative according to the standard (iterative process until acceptance; acceptance of responsible of monitoring the agreed tasks).

- **Phase F7 (Reflection on the problem and the future potential problems):** Closure of the project—the team is released from the responsibility, evaluating the obtained results (project goals, objectives, operational knowledge), and the department accepts and acknowledges the work done (Figure 7.18).

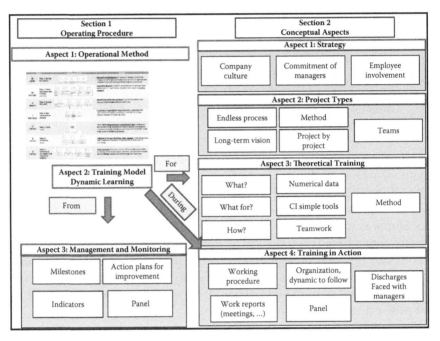

Figure 7.18 Continuous improvement sustainable model. (Eguren et al. (2010) Model/Framework for continuous improvement programme development to gain sustainable performance improvement in manufacturing facilities: an empirical study. With permission.)

CONTINUOUS IMPROVEMENT PROGRAMME IN A CASTING FACILITY

This medium-sized manufacturing plant, which is a premier supplier in the automotive industry, dedicates to machining cast iron brake disks. In order to reduce defective disc hub, they have applied dynamic structured problem solving by improvement teams where everyone is involved, from top management to production workers. The evaluation of the model leaded by researchers involved in the project shows that this project has been implemented very effectively and efficiently, and that regarding the rest of ratios, team/basic learning, project organization, team and corporate culture, it can be confirmed that strong links between these factors and the theoretical training have been found.

References

Arica, E., and Powel D. J. (2010) ICT Integration for automatic real-time production planning and control. Paper presented at the proceedings of APMS Conference, Cuomo, Italy.

Baranek, A., Hua Tan, K., and Debnar, R. (2010) Knowledge dimensions of lean: Implications on manufacturing transfer. Paper presented at the proceedings of 18th EurOMA Conference, Oporto, Portugal.

Bateman, N. (2005) Sustainability: The elusive element of process improvement. *International Journal of Operation & Production Management* 25 (3-4): 261–276.

Bateman, N., and Arthur, D. (2002) Process improvement programmes: A model for assessing sustainability. *International Journal of Operations & Production Management* 22 (5): 515–526.

Bateman, N., and Rich, N. (2003) Companies' perceptions of inhibitors and enablers for process improvement activities. *International Journal of Operations & Production Management* 23 (2): 185–199.

Bessant, J., and Caffyn, S. (1997) High-involvement innovation through continuous improvement. *International Journal of Technology Management* 14 (1): 7–28.

Bessant, J.,. Caffyn, S., and Gallagher, M. (2001) An evolutionary model of continuous improvement behaviour. *Technovation* 21 (2): 67–77.

Bowler, M., and Kurfess, T. (2010) Retrofitting Lean manufacturing to current semiautomated production lines. Paper presented at the proceedings of APMS, Cuomo, Italy.

Christopher, M., and Towill, D. R. (2000) Supply chain migration from lean and functional to agile and customized. *International Journal of Supply Chain Management* 5 (4): 206–213.

Corti, D., Egaña, M. M., and Errasti, A. (2008) Challenges for off-shore operations. Findings from a comparative multi-case study analysis of Italian and Spanish companies. EurOMA Conference.

Crosby, B. P. (1979) *Quality is free.* McGraw-Hill, New York.

Curry, A., and Kadasah, N. (2002) Focusing on key elements of TQM evaluation sustainability. *The TQM Magazine* 14 (4): 207–216.

Deming, W. E. (1986) *Out of the crisis.* MIT Press, Cambridge, MA.

Doumeingts, G. (1984) Methode GRAI: Methode de conception des systems en productique. Thése d'état: Automatique: Université de Bordeaux 1.

Errasti, A., and Egaña, M. M. (2008) *Internacionalización de Operaciones productivas: Estudio Delphi.* CIL S01, San Sebastian, España.

Eguren, J. A., Goti, A., and Pozueta, L. (2011) Diseño, aplicación y evaluación de un modelo de Mejora Continua, *DYNA Ingeneria e Industria,* Feb.

Eguren, J.A., Goti, A., Pozueta, L., and Jaca, C. (2010) Model/Framework for Continuous Improvement Programme development to gain sustainable performance improvement in manufacturing facilities: an empirical study. *APMS International Conference*, 1 (1): 56–56.

Eguren, J.A., Pozueta, L. and Goti, A. (2010) Diseño y aplicación de un sistema de evaluación de un Modelo de Mejora Continua en una empresa auxiliar del automoción. 4th International Conference on Industrial Engineering and Industrial Management. XIV Congreso de Ingeniería de Organización. 1(1): 938–947.

Feigenbaum, A. V. (1986) *Control total de la calidad*. CECSA, Mexico.

Garcia-Sabater, J. J., and Martin-Garcia, J. A. (2009) Facilitadores y barreras para la sostenibilidad de la Mejora Continua: Un estudio cualitativo en proveedores del automóvil de la Comunidad Valenciana. *Intangible Capital* 5 (2): 183–209.

Goh, T. N. (2002) A strategic assessment of six sigma. *Quality and Reliability Engineering International* 18: 403–410.

Harrison, A., and van Hoek, R. (2005) *Logistics management and strategy*. Prentice Hall/Financial Time, Harlow, U.K.

Hoekstra, S., and Romme, J. (1992) *Integrated logistics structures: Developing customer oriented goods flow*. McGraw-Hill, London.

Hoerl, R. (2001) Six sigma black belts: What do they need to know? *Journal of Quality Technology* 33 (4): 391–406.

Hyland, P. W., Mellor, R., and Sloan, T. (2007) Performance measurement and continuous improvement: Are they linked to manufacturing strategy? *International Journal and Technology Management* 37 (3/4): 237–246.

Ishiwata, J. (1997) *Productivity through process analysis*. Productivity Press, Cambridge, MA.

Jaca, C. (2011) Modelo de sostenibilidad del trabajo en equipos de mejora, PhD discussion, San Sebastián, Tecnun, University of Navarra, Spain.

Jaca, C., Mateo, R., Tanco, M., Viles, E., and Santos, J. (2010) Sostenibilidad de los sistemas de mejora continua en la industria: Encuesta en la CAV y Navarra. *Intangible Capital* 6 (1): 51–77.

Jones, D. T., Hines, P., and Rich, N. (1997) Lean logistics. *International Journal of Physical Distribution and Logistics Management* 27 (3/4): 153–173.

Jørgensen, F., Boer, H., and Gertsen, F. (2003) Jump-starting continuous improvement through self-assessment. *International Journal of Operations & Production Management* 23 (10): 1260–1278.

Juran, J. M. (1990) *Juran y la planificación para la Calidad*. Diaz de Santos, Madrid.

Kolb, D. (1984) *Experimental learning: Experience as the source of learning and development*. Prentice-Hall, Upper Saddle River, NJ.

Kotter, J. P. (2007) *Al frente del cambio*. Empresa Activa, Barcelona.

Lyonnet, B., Pralus, M., and Pillet, M. (2010) Critical analysis of a flow optimization methodology by value stream mapping. Paper presented at the proceedings of APMS, Cuomo, Italy.

Magnusson, M. G., and Vinciguerra, E. (2008) Kay factors in small group improvement work: An empirical study at SKF. *International Journal Technology Management* 44 (3): 324–337.

Martinez, S., Errasti, A., and Eguren, J.A. (2012) Lean–Six Sigma approach put into practice in an empirical study. XVI Congreso de Imgeniena de Organizacion, Ngo, Spain.

Mason-Jones, R., Naylor, B., and Towill, D. (2000) Engineering the agile supply chain. *International Journal of Agile Management Systems* 2 (1): 54–61.

Middel, R., Gieskes, J., and Fisscher, O. (2005) Driving collaborative improvement processes. *Product Planning and Control* 16 (4): 368–377.

Nakajima, S. (1998) *An introduction to TPM*. Productivity Press, Portland, OR.

Nakano, M. (2010) A concept for lean manufacturing enterprises. Paper presented at the proceedings of APMS, Cuomo, Italy.

Nonaka, I. (1994) A dynamic theory of organizational knowledge creation. *Organization Science* 5(1).

Ohno, T. (1998) *Toyota Production System: Beyond large-scale production*. Productivity Press, Cambridge, MA.

Olhager, J. (2003) Strategic positioning of the order penetration point. *International Journal of Production Economics* 85 (3): 319–329

Pozueta, L., Eguren, J A., and Elorza, U. (2011) The "factory of problems": Improvement of the quality improvement process. Paper presented at the proceedings of the 14th QMOD Conference on Quality and Service Sciences, , University of Navarra, Technun, San Sebastian, Spain, pp. 1439.

Prajogo, D. I., and Sohal, A. S. (2004) The sustainability and evolution of quality improvement programmes: An Australian case study. *Total Quality Management & Business Excellence* 15 (2): 205.

Pyzdek, T. (2003) *The six sigma handbook: A complete guide for green belts, black belts, and managers at all levels*, 2nd ed. McGraw-Hill, New York.

Ruggles, R. (1998) The state of the notion: Knowledge management in practice. *California Management Review* 40 (3).

Schroeder, D. M., and Robinson, A. G. (1991) *America's most successful export to Japan: Continuous improvement programs*. MIT Sloan, Cambridge, MA.

Schroeder, R. G., Linderman, K., Liedtke, C., and Choo, A. S. (2007) Six sigma: Definition and underlying theory. *Journal of Operations Management* 26: 536–554.

Senge, P. M. (2005) *La quinta disciplina. El arte y la práctica de la organización abierta al aprendizaje*, 2nd ed. Granica, Buenos Aires.

Szeto, A. Y. T., and Tsang, A. H. C. (2005) Antecedents to successful implementation of six sigma. *International Journal of Six Sigma and Competitive Advantage* 1 (3): 307–322.

Szulanski, G. (1996) Exploring internal stickiness: Impediments to the transfer of best practices within the firm. *Strategic Management Journal* 17.

Szulanski, G. (2003) *Sticky knowledge: Barriers to knowing in the firm*. Sage Publications, London.

Taylor, D., and Brunt, D. (2010) *Manufacturing operations and supply chain management: The lean approach*. Cengage Learning, Andover, U.K.

Tort-Martorell, J., Grima, P., and Marco, L. (2008) Sustainable improvement: Six sigma lessons after five years of training and consulting. *Corporate Sustainability as a Challenge for Comprehensive Management* 57–66.

Turbide, D. (1998) APS: Advanced planning systems. *APS Magazine*.

Upton, D. (1996) Mechanisms for building and sustaining operations improvement. *European Management Journal* 14 (3): 215–228.

Upton, D., and Bowon, K. (1998) Alternative methods of learning and process improvement in manufacturing. *Journal of Operations Management* 1–20.

Wilkner, J., and Rudberg, M. (2005) Integrating production and engineering perspectives on the customer order decoupling point. *International Journal of Operations and Production Management* 25 (7): 623–640.

Womack J., and Jones, D. (2005) *Lean thinking*. Free Press, New York.

Wu, C. W., and Chen, C. L. (2006) An integrated structural model toward successful continuous improvement activity. *Technovation* 26: 697–707.

Further Reading

Chen, K., and Ji, P. (2007) A mixed integer programming model for advanced planning and scheduling (APS). *European Journal of Operations Research* 181: 515–522.

De Haan, J., and Overboom, M. (2010) Mass, lean and agile production: What is in a name? Paper presented at the 18th Proceedings of EurOMA, Oporto, Portugal.

Imai, M. (1986) *Manufacturing strategy: The strategic management of the manufacturing functions*. Macmillan Education, London.

Jaca, C., Viles, E., Ricardo, M., and Santos, J. (2012) Components of sustainable improvement systems: Theory and practice. *The TQM Journal* 24 (2): 142–154.

Nonaka, I. (1991) The knowledge creating company. *Harvard Business Review* November–December.

Slack, N., Chambers, S. and Johnston, R. (2004) *Operations management*, 4th ed. Prentice Hall/Financial Time, Harlow, U.K.

Sohal, A., Olhager, J., Neil, O., and Prajogo, D. (2010) Implementation of OEE-issues and challenges. Paper presented at the Proceedings of APMS, Cuomo, Italy.

van Hoek, R. (1998) Reconfiguring the supply chain to implement postponed manufacturing. *International Journal of Logistics Management* 9 (1).

chapter 8

Supplier Network Design and Sourcing

Ander Errasti

> *It is common sense to take a method and try it. If it fails, admit it frankly and try another. But above all, try something.*

> **Franklin D. Roosevelt**

Contents

Introduction

In this chapter, we discuss:

- Total cost ownership and purchasing functions within the value chain
- Global purchasing

- Purchasing procedures
- Procurement and sourcing

Note: To look for more deep analysis, the author recommends another book about industrial purchasing management (Errasti, 2012).

Total Cost Ownership

Purchasing and Procurement Functions

Purchasing has evolved in the past few decades from a passive administrative role into a strategic function that contributes to creating a competitive advantage as much as other business functions (Alinaghian and Aghadasi, 2006). This development is logical, given that purchases represent a large percentage of the final cost of the product and are of crucial importance for its quality and performance. The shift from a traditional administrative transaction to an innovative relationship with suppliers has followed a trajectory characterized by two main phases:

1. An initial phase of better performance of the logistical functions under production philosophies, such as Just-in-Time (JIT) and other continuous improvement management philosophies, such as Total Quality Management (TQM). This phase started with the breakthrough in the paradigm that suppliers are part of the extended enterprise (Childe, 1998).
2. A second phase, where purchasing management has a stage evolving from an operations function to a strategic level function, where decisions are made with the aim of supporting a company's sustainability advantage. There are purchasing policies that take advantage of complex market situations and agreements with suppliers at strategic levels characterized by co-design activities for the development of products and even advanced supply chain practices.

Thus, the **new purchasing function** contributions include:

- Define the strategy in the supplier market.
- Search for new suppliers.
- Collect suppliers' value-added proposals and **develop** appropriate **suppliers.**
- Integrate marketing and supply strategies in shorter product life cycles.
- Define the **contract management** in the contract life cycle.
- Relate **supplier management** to quality assurance in new product development.
- Reduce total cost ownership.

Traditional procurement functions seek to guarantee the **service level** of suppliers (selected according to the purchasing function) in order to meet the company's operations needs with **less working capital** and **minor management costs**.

For this purpose, procurement managers must ensure:

- **Demand forecasting information** is sent to **critical suppliers** in order to guarantee long-term supplies and assure suppliers' production availability.
- **Planning and scheduling and rescheduling** of supply needs are subordinated to production planning and scheduling.
- The mode of **transport form** is chosen and the incoterms (international commerce terms) suggested by the purchasing function are fixed.
- Stock management.
- Product receiving and put away.
- Maximally efficient line **production/assembly supply** (Kanban, replenishment, just-in-sequence).

Cheap versus Economic Purchase

Total cost of ownership is a methodology and philosophy that looks beyond the price of a purchase to include many other purchase-related costs, such as procurement and quality. This approach has become increasingly important as organizations look for ways to better understand and manage their costs (Ellram, 1995).

Purchasing function activities do not end once the contract agreement is settled. Sometimes better price conditions transform into worse costs.

The concept of total cost of ownership tries to diminish the variability of extra cost in the supplier selection phase by integrating product and service quality assurance procedures and monitoring the extra cost generated by suppliers when negotiating the next tender. Thus, the total cost of ownership models are then further classified by their primary usage: supplier selection or supplier evaluation.

This analysis could be appropriate when comparing local suppliers in a high-cost country and new suppliers in an offshore low-cost country in a global sourcing strategy or comparing new suppliers close to a new offshore facility and the current suppliers of the manufacturing network. As we can see in the following example, labor cost is an important factor, but to estimate the potential benefit correctly, other operations costs should be considered (Figure 8.1).

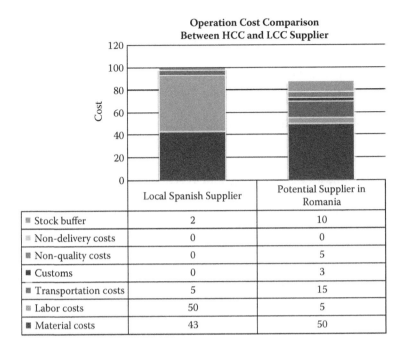

Figure 8.1 Comparing operations costs from local medium-cost country versus new offshore low-cost country supplier.

Global Purchasing

Purchasing Department

Some authors (Fung, 1999; Fant and Panizzolo, 2006) synthesize three different roles in a purchasing department:

- **Rationalization:** Purchasing contributes to the company's competitive strength by minimizing total costs of production, logistics, price, and nonquality costs.
- **Structure:** Refers to managing the company's supplier network, especially in terms of the degree of the company's dependence on specific suppliers or on managing improvement programs.
- **Development:** Concerns the alignment of business strategy and strategic product and process development.

Even though the initial purchasing functions are usually centralized in the matrix, the global operations managers have to consider how to develop the purchasing function in the new offshore facility's organization, especially when new supplier bases are needed in order to reduce total cost of ownership and better responsiveness or when new product development will take place in the new facility.

To develop this function, the requirements for purchasing managers and departments have four areas of management (Giannakis, 2004):

- **Product management:** Involves the activities performed to define, assess, and control the product/service attributes that are exchanged.
- **Procurement management:** Involves the activities to plan and control the delivery of products.
- **Contract management:** Besides the terms and conditions that specify the price and deliverables, contract management includes performance criteria, incentive negotiation and rebates, mutual safeguards, and other critical variables.
- **Supplier development:** Refers to cooperative efforts between a buying firm and its suppliers to ensure the efficiency and effectiveness of supplier operations.

The tasks that could be delegated in each purchase category (Giannakis, 2004) for these four management areas are shown in Table 8.1 to Table 8.3.

Sourcing Organization Types

The organization of purchasing departments should aim to maximize the overall efficiency and try to adapt to the needs of business units and their different implementations of research and development (R&D) and production worldwide. In order to achieve that, the purchasing organization should be considered to the family level or purchase category (Trautmann, Bals, and Harmann, 2009).

Centralized, Decentralized, and Hybrid Structures

The main reasons to centralize purchasing include the exploitation of economies of scale and the negotiation power with the suppliers (Mathyssens and Faes, 1997; Karjalainen, 2011).

The economy of scale and purchasing power refers to getting lower unit costs by adding volume to the different units and unifying this volume through the product standardization.

It is common to think about purchasing centralization of raw materials, nonproduction purchases, and supplies with a high degree of standardization or normalization.

Other advantages associated with the centralization are the possibility of buyers' specialization in both technical knowledge of the product, which enables them to achieve an advantage through a more accurate specification of the supplies they are buying, as well as market

Table 8.1 Product management and procurement tasks

	Precontractual	Institutional	Operational
Product management	1. Collect information on new goods being developed, or already available in supplier markets 2. Evaluate goods/services in terms of value to organization 3. Promote standardization and simplification of parts	4. Establish goods/services technical specifications 5. Establish goods/services design variations 6. Establish goods/services deliverables (materials approvals papers/product tests) and prototypes	7. Receive product/services 8. Product data management (data creation, transfer, etc.) 9. Suggest alternative products and technologies
Procurement management	10. Procurement process mapping (order delivery process) 11. Examine potential suppliers' processes to identify areas of expertise and areas that need improvement (e.g., forecasting and order information, transport form, packaging) 12. Select supply strategy to exchange the product/service (e.g., make to stock, assembly to order, make to order)	13. Establish the information exchange systems for order management (order omission, confirmation, delivery, invoice) 14. Establish the operations agreed with third party logistics 15. Establish means of transport and delivery (returnable unit loads, etc.)	16. Order main system and alternative/redundant systems 17. Process standardization of more efficient practices (e.g., extranet versus fax, exw versus ddu transport) 18. Measure process effectiveness (total logistic cost, service level, stock turnover) 19. Benchmark processes against targets 20. Audit suppliers

Source: Adapted from Giannakis, M. (2004) The role of purchasing in the management of supplier relationships. Paper presented at the 2004 EurOMA Operations and Global Competitiveness, Fontainebleau, France.

Table 8.2 Contract management tasks adapted

	Precontractual	Institutional	Operational
Contract management	**Determinants of relationships factors:** 21. Evaluate market/industry norms 22. Evaluate strategic choice of the company 23. Evaluate organizational culture 24. Evaluate requirements of customers 25. Explore external market and supplier information 26. Capture and transfer internal learning	**Contractual agreements:** 27. Select the legal form of the relationship (partnership, alliance) 28. Select the contract type 29. Select the length of the relationship 30. Establish contractual safeguards 31. Delegate role and responsibilities with suppliers and internal customers 32. Provide suppliers with financial incentives 33. Adapt relationships specific agreements **Noncontractual agreements:** 34. Make spirit of fellowship agreements 35. Provide unconditional help to suppliers 36. Use managerial expertise 37. Ensure supplier is regularly updated	**Monitor and control:** 38. Ensure clear contract deliverables 39. Define key performance indicators related to procurement 40. Define key performance indicators related to product quality **Authorize all appropriate spending for:** 41. Mandated levels of spending 42. Benchmark supplier performance as external market and contract criteria 43. Monitor procurement from other suppliers

Source: Adapted from Giannakis (2004) The role of purchasing in the management of supplier relationships. Paper presented at the 2004 EurOMA Operations and Global Competitiveness, Fontainebleau, France.

Table 8.3 Supplier management tasks adapted

	Precontractual	Institutional	Operational
Supplier management	**Supplier selection:** 44. Identify appropriate strategic and tactical issues 45. Develop and maintain supplier databases 46. Prepare request for quotations 47. Identify potential suppliers (preassessment) 48. Tender 49. Evaluate suppliers (reputation, capabilities, cost, etc.) 50. Evaluate business cost, supplier risk, exit strategy 51. Define supplier approval (screening) 52. Negotiate terms of contract 53. Define supplier certification **Supplier development:** 54. Identify critical processes/product development 55. Develop a cross-functional supplier development team	56. Identify critical suppliers to develop 57. Provide resources, methodologies to make the diagnose from critical suppliers 58. Establish improvement programs and organize workshops to develop reengineering projects 59. Establish recognition and reward	60. Monitor the improvement programs and projects with suppliers 61. Evaluate the supplier development program

Source: Adapted from Giannakis, M. (2004) The role of purchasing in the management of supplier relationships. Paper presented at the 2004 EurOMA Operations and Global Competitiveness, Fontainebleau, France.

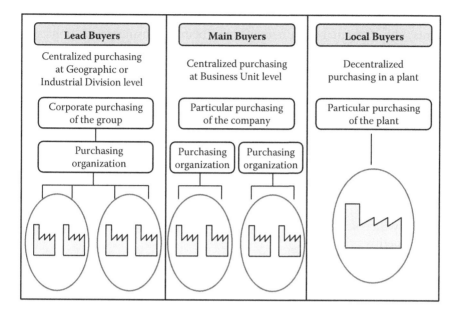

Figure 8.2 Degree of purchase centralization.

specialization, which allows them to search suppliers on a global level (McCue and Pitzer, 2000).

This degree of centralization can be at a business unit level or at a geographic level of an industrial division or corporation. Depending on the level, there are different purchasing managers (Gelderman and Semeijn, 2006) (Figure 8.2):

- **Lead buyers:** Purchasing managers that manage centralized purchasing at a geographic or industrial division level.
- **Main buyers:** Purchasing managers that manage centralized purchasing at a business unit level.
- **Local buyers:** Purchasing managers that manage decentralized purchasing in a plant.

On the other hand, decentralized purchasing allows for a better understanding of local requirements not only in terms of cost, but also of quality and service. Moreover, this type of organization gets a greater reactivity when purchasing management requires a quick response (e.g., process of new product development) (Van Weele and Rozemeijer, 1996; Hult and Nichols, 1999).

A particular case of decentralization is the project purchaser. Here the appointee is responsible for the purchase associated with the project during the project life cycle.

It is normal to consider the purchasing decentralization of components with particular requirements for a plant or business unit, nonstandardized quality requirements, and demand in delivery time.

Nevertheless, organizations often choose a hybrid model, with practices differing, for example, by product category. In hybrid purchasing organizations, there is a division of tasks between the head office and local site. For example, the head office takes responsibility for the negotiation of some long-term contracts and the subsidiaries issue orders against these contracts (Trautmann, Bals, and Harmann, 2009).

Supplier Market Analysis

In a manufacturing company, there usually are different product categories for which suppliers have different market conditions, as determined by competitive rivalry.

The purchasing functions have to be prioritized into which categories managers will concentrate their improvement efforts. For this purpose, the matrix in Figure 8.3 could be appropriate. The matrix is an ABC analysis where the criteria to select the A (or first wave) are the potential savings and implementation difficulty.

Purchasing Policies

Purchasing policies depend on the potential risk/benefit of the purchase decision and the power relation between suppliers and purchasers.

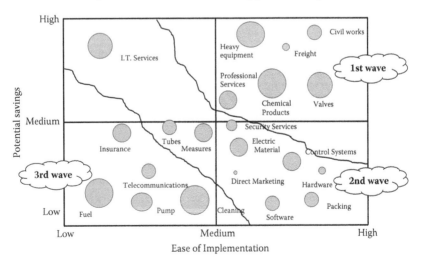

Figure 8.3 Potential savings and implementation difficulties for different purchase categories.

The factors to consider in determining the potential risk/benefit of the purchasing decision include:

- **Cost of nondelivery.** The supply interruption probability due to logistical routes, social or economic problems for the supplier, or quality failure.
- **Benefits due to logistical,** quality, and information and communication technology (ICT) integration.
- Benefits due to better product specification or newly developed products.

The factors to consider in the power relation between supplier and purchaser include:

- **Supplier market rivalry** (uniqueness of product and technology, market demand versus supplier capacity, number of equivalent suppliers, concentration of suppliers or new suppliers).
- **Supplier purchaser power** (purchase volume, cost to change supplier in terms of capital and know-how entry barriers with the same product, cost to change product in terms of capital and know-how entry barriers, cost to change supplier in terms of logistics route).

These two groups of factors determine which possible purchasing policy (partnership, cooperation, imposing, trading, and opportunism) could be more effective in each case (Figure 8.4).

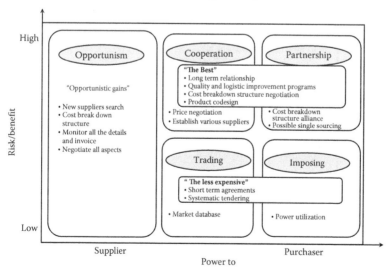

Figure 8.4 Most effective purchasing policies related to risk/benefit and power to purchaser of supplier analysis.

Note: For further reading to develop partnerships, see Errasti et al., (2007).

Kraljic (1983) proposed another matrix for mapping and defining the most appropriate purchase policies. This approach takes into account the **importance of purchasing and the complexity of the supply market**. The strategic importance of purchasing in terms of value-added by a product line, the percentage of raw materials in the total cost of the final product, as well as their impact on profitability and the complexity of the supply market in terms of supply scarcity, the pace of technology and/or materials substitution, entry barriers, logistics cost, complexity, and monopoly or oligopoly conditions make purchasing a key issue in many companies. By assessing the company's situation in terms of these variables, the supply strategy could be determined by trying to exploit purchasing potential and diminish risks.

There are **four purchasing item classes**—(1) leverage items or volume suppliers, (2) strategic items, (3) noncritical items, and (4) bottlenecks or midcritical items—for which supply strategies are proposed (Figure 8.5).

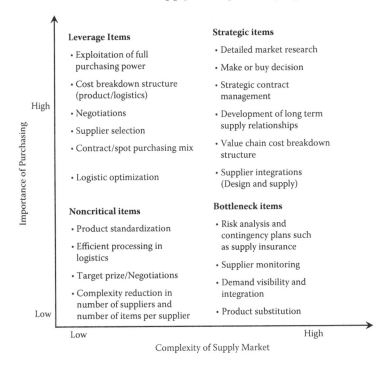

Figure 8.5 Importance of purchasing and complexity of supply market matrix. (Adapted from Kraljic, P. (1983) Purchasing must become supply management. *Harvard Business Review.* 61(5), 109–117.)

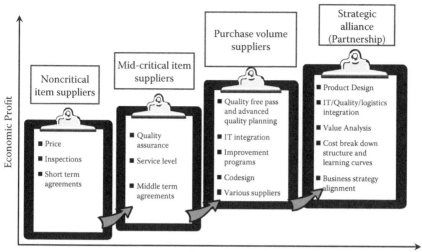

Figure 8.6 Best supplier development practices for noncritical suppliers, midcritical suppliers, purchase volume suppliers, and strategic suppliers.

Taking into account a supplier's maturity level, the purchasing policies outlined in Figure 8.5 could be put into a matrix of maturity development and economic profit (Figure 8.6).

Purchasing Tools and Techniques

The alternatives to improving purchasing management could be divided into market- and technical-driven initiatives (Figure 8.7).

Market-driven initiatives try to exploit purchasing volume with three main techniques:

- Concentrate purchasing volume and create competence
- Evaluate best price monitoring and negotiating
- Global sourcing and create new rules by searching or creating new alternative suppliers

Technical-driven initiatives try to adapt product needs more accurately, create an efficient supply chain, and establish strategic relationships. Also there are three main techniques:

- Improve product technical specifications to needs
- Improve processes and integration in the extended enterprise
- Restructure relationships with suppliers

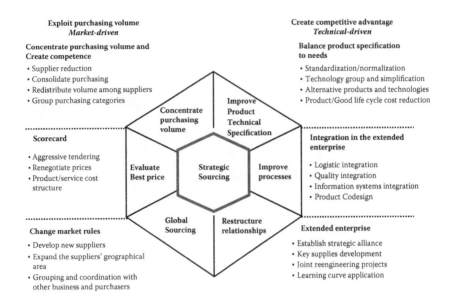

Exploit purchasing volume
Market-driven

**Concentrate purchasing volume and
Create competence**

• Supplier reduction
• Consolidate purchasing
• Redistribute volume among suppliers
• Group purchasing categories

Scorecard

• Aggressive tendering
• Renegotiate prices
• Product/service cost
  structure

Change market rules

• Develop new suppliers
• Expand the suppliers' geographical
  area
• Grouping and coordination with
  other business and purchasers

Create competitive advantage
Technical-driven

**Balance product specification
to needs**

• Standardization/normalization
• Technology group and simplification
• Alternative products and technologies
• Product/Good life cycle cost reduction

**Integration in the extended
enterprise**

• Logistic integration
• Quality integration
• Information systems integration
• Product Codesign

Extended enterprise

• Establish strategic alliance
• Key supplies development
• Joint reengineering projects
• Learning curve application

Concentrate purchasing volume — Improve Product Technical Specification — Evaluate Best price — Strategic Sourcing — Improve processes — Global Sourcing — Restructure relationships

Figure 8.7 Purchasing management techniques.

Purchasing Management and Product Life Cycle

The applicable purchasing techniques and their effectiveness in total cost of ownership reduction depends on the product and the supplier life cycle (Figure 8.8).

Purchasing Procedures

A purchasing procedure has to be defined, standardized, and established if there is a need to develop new suppliers, led by the offshore facility (Figure 8.9).

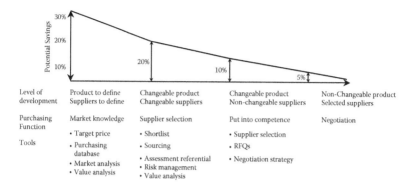

Figure 8.8 Potential savings and product supplier life cycle.

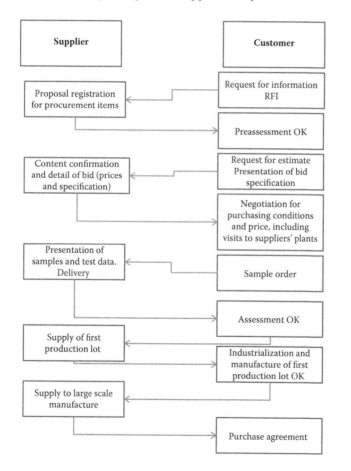

Figure 8.9 Purchasing procedure for new supplier development for manufacturing companies.

Case: Ternua–Astore

HOW SMES COMPETE IN THE OUTDOOR AND SPORT TEXTILE INDUSTRY WITH THE MULTINATIONALS

The clothing industry is one of the most dynamic retailer-driven and global economic sectors. This industry is characterized by price sensitive customers, short product life cycles, a wide product range, as well as volatile and unpredictable demand. There are three types of clothing retailers: leading brand retailers (e.g., Benetton, Zara, H&M, GAP), hypermarket brands for supermarket retailers (e.g., Walmart), **value brand retailers** (e.g., Desigual). Despite the characteristics mentioned above, the traditional configuration is a forecast driven supply chain distributing two campaigns; Spring-Summer and Fall-Winter.

The effectiveness of this approach has been questioned because of: difficulties to get collections in time, on full and error free; overstock after campaigns due to forecast driven manufacturing and procurement and inaccurate demand forecast and out of stock for garments with better than forecasted sales for the campaign or season. However, the Spanish clothier Zara broke this paradigm developing a super-responsive and quick supply chain for leading brands which is called "Pronto Moda" or Rapid-Fire Fulfillment.

Ternua and Astore are value brand retailers devoted to outdoor, urban and team wear clothing which are located in the Basque Country (North of Spain). They strive to offer the latest in garments with the goal of providing greater comfort and improving performance during strenuous physical activity. These companies have been operating a traditional supply chain configuration for a value brand retailer: Owned design and development team, high range of collections in terms of width and depth, a vertically disintegrated manufacturing process, long lead times due to that the manufacturing process is located far away, a disintegrated supply chain (fabric and manufacturing process) and multichannel distribution. The companies used to develop two campaigns (Spring-Summer and Fall-Winter) with a wide variety of garments per collection based on forecasted demand. The main inefficiencies with the current supply chain structure and management were:

- A wide range and variety of garments per collection that was not perceived necessary by the channel distributors and wholesalers.

- Overstock when finishing the campaign due to Minimum Order Quantity (MOQ) in production.
- Overstock due to a forecast driven supply chain. The orders to manufacturers should be settled before the sales convention, due to long lead times of fabric and trims procurement, as well as the fact that they have garments manufacturing in overseas facilities. This problem was larger for premium garments, which item costs were much higher.
- No possibility to procure and manufacture a second order for the campaign in garments that were selling in higher quantities than forecasted.

These deficiencies lead the companies to develop a new business strategy based on the following two objectives; accomplishing profitable campaigns and increasing sales by introducing more campaigns ("mini-collections") between the two major ones. The actions that were taken to mitigate the deficiencies and to change the supply chain structure in a profitable way were the following:

- By applying lean principles the companies managed to relaxing design resources and eliminate waste in new collections development within a campaign.
- By categorizing garments according to demand behaviour (continuity, season) and prices. The former to create a new category called "fresh", which means that these garments have a dynamic assortment, and the latter to establish three price levels (basic, medium and premium).
- Employing concurrent engineering among designers, purchasers, manufacturers and forwarders to reduce product development lead time, and the establishment of a critical chain and multi-project environment in order to avoid resource design constraints and resource allocation constraints for the manufacturing suppliers.
- The development of a global purchasing strategy and also, a global and local supplier network taking into account price position, assortment rotation and demand pattern.

In order to analyze how more campaigns could be introduced to increase the competitiveness of the company, the product portfolio was analyzed and segmented based on demand pattern (customer buying behavior) and price levels. The analysis resulted in six product categories; basic and continuity, premium and continuity, basic

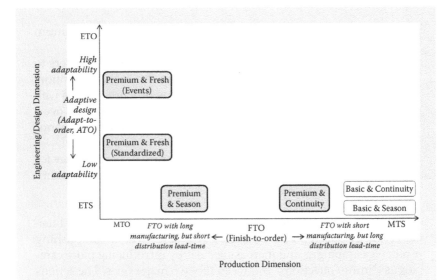

Figure 8.10 Product segment classification in the new supply chain strategy.

and season, premium and season, premium and fresh (standardized), and premium and fresh (events). But the changes were only necessary in fourth of them. The main changes in the supply chain strategy were (see Figure 8.10):

- **Premium and continuity:** CODP (customer order decoupling point) was moved upstream in the supply chain to be able to better react to new customer demand and to be able to issue replenishment orders, should the demand be higher than forecasted. Therefore the new supply strategy is finish-to-order (FTO) with the CODP at the end of the fabric sourcing cycle with production kept in low-cost countries. Lean principles have been applied to reduce the manufacturing lead-time, but production in low cost countries still makes the distribution lead-time rather lengthy.
- **Premium and season:** similar changes were made to this segment with the difference in that production is kept in medium cost countries to reduce replenishment lead-times even further and thereby allowing for quick response replenishment. The focus was on agile manufacturing in order to achieve flexibility in terms of volume and time.
- **Premium and fresh (standardized):** is ATO/MTO with limited adaptability in design. Operations are concentrated to medium

cost countries applying agile manufacturing in order to reach flexibility, and to keep lead-times within acceptable limits.
- **Premium and fresh (events):** is ATO/MTO with a high level of adaptability in garment design. Design and operations are typically located in high and medium cost countries applying agile manufacturing to get flexibility and quick response, both in terms of changes in design and in customer demand.

The result of the changes in the product segments are summarized in Figure 8.11, presenting the percentage of total demand against percentage of products. It can see that the products segments modified are 65% of the total demand. The products to the right respond to demand with an agile supply chain (operations closer to the major markets mainly focusing on responsiveness), whereas the products to the left respond with a lean supply chain (majority of operations in low-cost countries). It has been also considered the two dimensions of the CODP (engineering and production) for the different product segments.

The new product segmentation and product strategy, combined with the newly developed supply chain strategy, offers not only

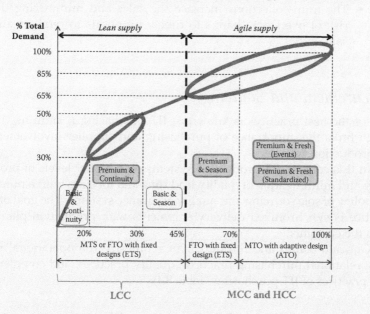

Figure 8.11 The six product segments distributed along a lean/agile and total demand continuum.

shorter lead-times for all product segments, but also alternative supply chains allowing for mini-collections with very short design and
delivery lead-times. The former using a postponement strategy in
low cost countries, and the latter using medium or high-cost country suppliers located in close proximity to the main markets to keep
lead-times at a minimum.

The experienced and projected results indicate the following:

- Campaigns could be more profitable if the collections are segmented taking into account the demand characteristics (continuity, season, etc.) and prices, establishing the most adequate
  supply strategy for each respective segment.
- Improving assortment rotations requires that the new product
  development and order fulfillment processes are redesigned.
  The latter typically leading to reconfiguration of the supply
  chain and manufacturing operations.
- Introducing a product platform strategy facilitates concurrent
  engineering (designers, purchasers, manufacturers, forwarders), which in turn allows for reducing product development
  and order fulfillment lead times.
- The mini-collections increase the sales and minimizing the
  risk of overstock thanks to timely proposals to retailers and
  sales outlets.

Procurement and Sourcing

One of the best practices is known as JIT purchasing (Lamming, 1996),
which links the importance of purchasing and supplier involvement in
JIT production systems.

In this context, JIT production systems seek higher levels of productivity and minimization of quality lead time and lot sizes with a purchasing policy of sole sourcing and quality assurance systems. The goal of this
practice is synchronized delivery shipments with production planning
from the customer.

Gonzalo-Benito and Spring (2000) synthesize the operational practices, relational purchasing practices, quality practices, and complementary practices of JIT purchasing (Table 8.4).

Demand Visibility and Procurement Management

Beyond the boundaries of a single company, the flows between companies in a supply chain can be classified into three categories: material,

Table 8.4 Operational practices, relational purchasing practices, quality practices, and complementary practices

Operational Practices	Purchasing Relational Practices	Quality Practices	Complementary Practices
Frequent delivery (small batch size)	Risk sharing	Quality assurance certification	Supplier involvement in design and development
Reduced inventory	Vendor management inventory		Supplier development programs
Kanban suppliers	Vendor management inventory	Supplier selection based on quality and reliability	
Fulfillment of delivery in tight time windows		Time windows fulfillment development programs	
ICT integration (EDI, extranets, etc.)	Long-term relations		
Region/area sourcing	Single source agreement in an area/region		Business strategy alignment
Load unit standardization and returnable units	Cost based price calculating		

Source: Adapted from Gonzalo-Benito, J., and Spring, M. (2000) JIT purchasing in the Spanish auto components industry: Implementation patterns and perceived benefits. *International Journal of Operations and Production Management* 20 (9):1038–1061.

information, and decision. Material flows are related to raw material, components, and product supply. Examples of information flow are bid requests, purchase orders, invoices, etc. **Decision flows are sometimes mistaken as information flows,** but decision flows are at the upper level of complexity. **The communication of decisions between companies is composed of information, but also via a negotiation process.** Taking a "purchase order" as an example of information flow, it can be transformed into a decision flow as a "demand plan." The customer provides its supplier with a demand plan with a defined horizon (e.g., two months) and periods (e.g., weeks) at an agreed time interval (e.g., two weeks). The supplier analyzes the demand plan, regards its capacity and supply constraints, and negotiates some changes with the customer with the aim of achieving an efficient production and supply process. **A win–win negotiation between companies begins a collaborative network.** In such a network, the companies believe that together they can achieve goals that would not be possible if each one attempted them individually. Collaboration is a powerful instrument for increasing the efficiency of the entire network. Among the different mechanisms for managing collaboration between companies, collaborative decision-making modeling is one of the most valuable. In order to build collaborative networks, several challenges must be faced, such as trust, culture change, interoperability, etc., and the decision system modeling, which identifies the intracompany and intercompany decision flows is a good starting point for reaching success. The GRAI model has been successfully used to model, analyze, and improve the decision system of a single company, in addition to supply chains and complex enterprise networks. By modeling the entire decision system of a supply chain, all the decision activities, the decision flows, and the feedbacks are identified, and the decision system can be improved by means of synchronizing and removing inefficiencies.

There are **two new problems with procurement management** that have to do with **long lead times.** On the one hand, Western facilities are searching for new suppliers in low-cost countries, and, on the other hand, there are Western manufacturing companies that have located a new facility offshore, but they have not developed a local base of suppliers.

Thus, for these two cases, there is a need to guarantee the service level, but in some parts or components that were originally managed with material requirement planning linked to the facility planning and scheduling systems. Nevertheless, this is not possible anymore because the planning horizon of the facility is shorter than the transportation time between the two points. Thus, there is a need to manage the stock and replenishment of parts (Errasti et al., 2010) whose demand has to be forecasted and whose supplies could be affected by transport time reliability. In conclusion, **there is a need to change from JIT to Just-in-Case management (Figure 8.12).**

In this case, companies have to develop a system for forecasting item demand by opening the fixed planning horizon and they have to develop graphic tools to **dynamically monitor the stock position,** taking into account real demand variation, supply reception time, stock in transit, and orders delivered. In conclusion, **an extended** material requirements planning **(MRP) based partly on a forecast demand technique is suggested.**

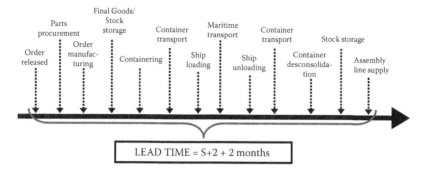

Figure 8.12 The transit time activities force companies to forecast demand, which opens the fixed planning horizon two months prior for suppliers who are not located nearby.

Figure 8.13 shows an example of an extended MRP based partly on forecast demand. The different stock levels represent:

- Light gray: Overstock → Stock level bigger than the maximum stock fixed by the financial department
- Dark gray: Warning → Stock level lower than security stock
- Black: Stock out → Forecast stock level less than 0

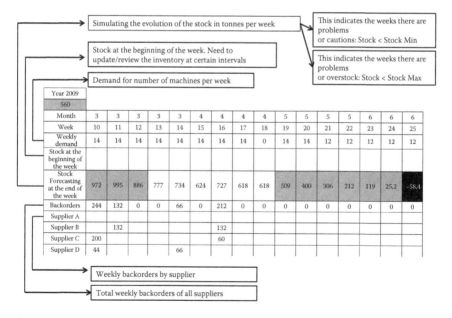

Figure 8.13 Extended MRP based in partly forecast demand.

References

Alinaghian, L. S., and Aghadasi. M. (2006) Proposing a model for purchasing system transformation. Paper presented at the EurOMA 16th Conference, Glasgow, Scotland.

Childe, S. J. (1998) The extended enterprise: A concept of cooperation. *Production Planning and Control* 9 (4): 320–327.

Ellram, L. M. (1995) TCO: An analysis approach for purchasing. *IJPD&LM* 25 (8): 4–23.

Errasti, A. (2012) *Gestión de compras en la empresa*. Ediciones pirámide, Grupo Amaya, Madrid, Spain.

Errasti, A., Chackelson, C., and Poler, R. (2010) An expert system for inventory replenishment optimization. In *Balanced Automation Systems for Future Manufacturing Networks*, Ortiz Bas, Á, Franco, R. D., Gómez Gasquet, P. (eds.), 129–136.

Fant, D., and Panizzolo, R. (2006) Purchasing performance measurement systems: A framework for comparison and analysis. Paper presented at the EurOMA 16th Conference, Glasgow, Scotland.

Fung, P. (1999) Managing purchasing in a supply chain context: Evolution and resolution. *Logistics Information Management* 12 (5): 362–366.

Gelderman, C. J., and Semeijn, J. (2006) Managing the global supply base through purchasing portfolio management. *Journal of Purchasing and Supply Management* 12: 209–217.

Giannakis, M. (2004) The role of purchasing in the management of supplier relationships. Paper presented at the EurOMA 2004 Operations and Global Competitiveness Conference, Fontainebleau, France.

Gonzalez-Benito, J., and Spring, M (2000) JIT purchasing in the Spanish auto components industry: Implementation patterns and perceived benefits. *International Journal of Operations and Production Management* 20 (9): 1038-1061.

Hult, G. T. M., and Nichols, E. (1999) A study of team orientation in global purchasing. *Journal of Business and Industrial Marketing* 14 (3).

Karjalainen, K. (2011) Estimating the cost effects of purchasing centralization-empirical evidence from framework agreements in the public sector. *Journal of Purchasing and Supply Management* 17: 87–97.

Kraljic, P. (1983) Purchasing must become supply management. *Harvard Business Review*, 61 (5): 109–117.

Lamming, R. (1996) Squaring lean supply with supply chain management. *International Journal of Operations and Production Management* 16 (2): 183–196.

Matthyssens, P., and Faes, W. (1997) Coordinating purchasing: Strategic and organizational issues. In *Relationships and networks in international markets*, eds. T. Ritter, H. G. Gemünden, and A. Walter (pp. 323–342). Elsevier, Oxford, U.K.

McCue, C., and Pitzer, J. (2000) Centralized vs. decentralized purchasing: Current trends in governmental procurement practices. *Journal of Public Budgeting, Accounting, & Financial Management* 12 (3): 400–420.

Trautman, G., Bals, L., and Harmann, E. (2009) Global sourcing in integrated network structures: The case of hybrid purchasing organizations. *Journal of International Management* 15: 194–208.

Van Weele, A. J., and Rozemeijer, F. A. (1996) Revolution in purchasing: Building competitive power through proactive purchasing. *European Journal of Purchasing and Supply Management* 2: 153–160.

chapter 9

The Ramp-Up Process

Sandra Martínez

A good plan, violently executed now, is better than
a perfect plan next week.

*Good tactics can save even the worst strategy. Bad tactics
will destroy even the best strategy*

Gen. George S. Patton

Contents

Introduction

In this chapter, we discuss:

- Factory operations and equipment management
- Ram-up process management
- Ramp-up project preparation
- Product/process alternatives
- Contingency plan and robust design

Factory Operations and Equipment Management: The Planning Process

The **factory life cycle** starts with strategic factory planning. This is the stage at which the factory defines its location, strategic role, products and volumes, and supply chain configuration, and where it identifies the required resources. Parallel to the building and civil engineering infrastructure project, the layouts and the internal/external handling and logistics have to be established.

The next steps represent factory ramp-up and operation, including facility management. Few factories have integrated these processes throughout the entire factory life cycle. Nevertheless, some authors have already identified that there are certain economic sectors, such as the wind power market with its dynamic changes, which require quick production ramp-up and also ramp-downs in the entire supply chain. This kind of production system, which was developed by wind energy company The Switch and is known as **The Switch Model Factory**, can be quickly built and is easily adaptable (Kurttila, Shaw, and Helo, 2010) (Figure 9.1).

Matching production capacity to drastically changing demands is an ongoing struggle that requires that capacity and capability be adjusted in a sustainable way for the following aspects:

- Demand: Changes in production volumes include managing uncertainty in both ramp-up and ramp-down phases.
- Product mix: Changes in product types within the same production facility.
- Life cycle: Entrance of new product models into a production and supply network, even in the ramp-up process,

The Switch Model Factory embraces the capabilities of the reconfigurable manufacturing system (RMS) capabilities (see Chapter 6) of speed, reliability, adaptability, and scalability in order to increase flexibility through reutilizing the first assembling design process.

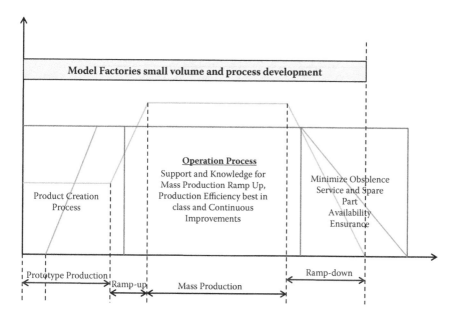

Figure 9.1. Product life cycle. (Adapted from Kurttila, P., et al. (2010) Model factory concept (from the vice president of supply chain of the company, The Switch, Finland: Enabler for quick manufacturing capacity ramp-up. Research paper.)

Ramp-Up Process Management

Ramp-up is a term used in economics and business to describe an increase in production ahead of anticipated increases in product demand. Alternatively, ramp-up **describes the period between product development and maximum capacity utilization**, which is characterized by product and process experimentation and improvements (Terwiesch and Bohn, 2001).

Some authors (T-Systems, 2010) point out that, strictly speaking, **ramp-up starts with the first unit produced and ends when the planned production volume is reached** (Figure 9.2). Nevertheless, in order to manage such a ramp-up with a high degree of precision, first of all a planning phase is necessary, starting with the engineering design of the product (desirable frozen design) or, in a more realistic way, with in-parallel product/process/network concurrent engineering (Figure 9.3) in project-based production systems (Errasti et al., 2008) or highly dynamic markets (Kurttila, Shaw, and Helo, 2010).

The ramp-up process should be a project-based activity where the requirements and needs of fast, efficient, and precise ramp-up are guaranteed by a **robust and resilient production system design** (Sheffi, 2006)

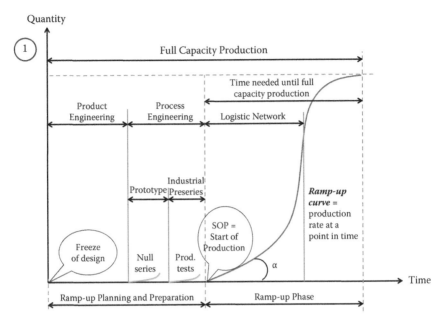

Figure 9.2 Traditional product design and process engineering and execution of ramp-up curve.

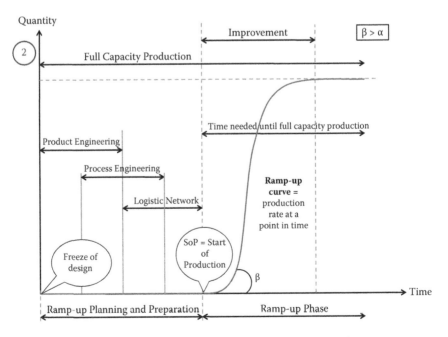

Figure 9.3 Concurrent design/process/network and execution of ramp-up curve.

that is characterized by the integration of the subsystems and functions. Ramp-up management requires monitoring measures that ensure quality and guarantee reasonable deviation in implementation time and cost.

One of the biggest problems a company faces during the ramp-up process is uncertainty. Many problems can occur that cause delay in time expectations. To explain the complexity of the problem, here's a brief description of an Asian manufacturer of filters:

In this company with 120 employees, there is a high level of control over interdepartmental information flow. This, added to low skill workers and a high absence rate, makes it reasonable to think that the ramp-up time for any project can be very long. To have an acceptable ramp-up process, it is necessary to impose a basic organization and define and structure the whole process and the work systems across departments in order to work efficiently.

If it is necessary for R & D to develop a project for a new range of filters, the engineer needs to have a minimum amount of information about all the issues related to the project (cost of materials, production process, customer concerns, etc.). Foresight is needed to know when the project is expected to end. However, if every decision made in developing projects or modernizing production is rejected due to high costs, the engineers in the R & D department end up being mere low-level employees who have no responsibility. As long as the company continues making profits, it is not really important to pay attention to shortening the ramp-up process or to make any other improvement.

Ramp-Up Project Preparation: A Facility's Physical Processes and Infrastructure

The typical scenario is one in which a new facility is implemented overseas. If we take into account the stages proposed in Chapter 7 for implementing the ramp-up process of the production management system and Greenfield's approach (Baranek, 2010), the ramp-up preparation (T-Systems, 2010) should take the following topics into consideration:

Plant and Factory Construction

It is often the case where new plants have to be built. Nevertheless, in **some cases**, the plants already exist and **the new production facilities have to be installed in an existing factory**. This is a common practice that reduces time to market/volume in sectors such as the automotive industry (Errasti et al., 2010).

Factory Layout and Material Flow

It is often useful to **simulate material flow**, including stocks and buffers, to learn about the behavior of the system under various conditions and **to evaluate** whether **the planned set-up is feasible, configurable, and adaptable** to different product/process scenarios (see Chapter 6).

Machine Procurement and Installation

If ramp-up includes a completely new setup of production equipment, the appropriate production technologies and machines have to be addressed. In cooperation with the production engineer, the available **vendors and machines in the market must be evaluated and rated** (see Chapter 8). Depending on the machines selected, a change in the engineering processes and/or the factory layout and material flow might become necessary (see Chapter 6).

Procurement of Tools and Jigs

Similarly, the appropriate tools have to be selected. Tailor-made tools, such as **stamping or casting dies, have to be watched because they are usually in the ramp-up project's critical path** and have an important role in the operational phase of the series production.

Detailed Workstation Design

The macro layout should be detailed for the different production areas and fit with the management system's ramp-up process (see Chapter 6).

Readjustment of Product and Production Processes

Product and production engineers should align their process specifications during the product engineering process; nevertheless, for different production processes, **there is a need to redesign the product** (dimensions, materials, etc.).

Sourcing, Supply Network, and Supply Chain

Logistics are one of the success factors of the ramp-up process, thus, the **sourcing policy and the logistic routes for local/global suppliers have to be addressed** (see Chapter 8).

Ramp-Up Support Processes

There are other processes that must take place during and after ramp-up. These processes have to be defined, tested, taught, and implemented as part of the ramp-up itself.

One of these processes is to establish the Advanced Quality Planning procedures and the predefined workflow and milestones in order to assure that few product process quality problems will occur. This is one of the key support processes due to the fact that most of the delays in facility implementation are caused by quality problems.

Another process is to write the product documentation and establish product variances, features, and configuration options. This configuration may change if there are no features or options because purchasing policies lead to changes in customers, individual parts, or materials.

Another process or aid for designing the ramp-up process is the Digital Factory process. The Digital Factory concept focuses on the integration of the methods and tools available at different levels to plan and test the product and the related production process from the early design phase to the operative control of the factory (VDI 4499, 2006). The Digital Factory-based methodology can even be applied in small and medium enterprises (SMEs) (Spath and Potinecke, 2005).

The Digital Factory (Figure 9.4) integrates the following processes:

- Product development, testing, and optimization
- Production process development and optimization
- Plant design and improvement
- Operative production planning and control

Alternative Product/Process Configurations in the Ramp-Up Process

Companies should expect production downtimes, ramp-up delays in time volume, and loss of production, especially when new factors, such as new products, new machinery, new suppliers, and new personnel, are introduced (Figure 9.5). Examples of delays are shown in Table 9.1.

Local/Global and Make versus Buy

The ramp-up strategy also may consider fragmenting production so that part of the process takes place in the new facility, supplying the rest from an existing one, or, alternatively, buying part of the production process and supplying it locally (see Chapter 4).

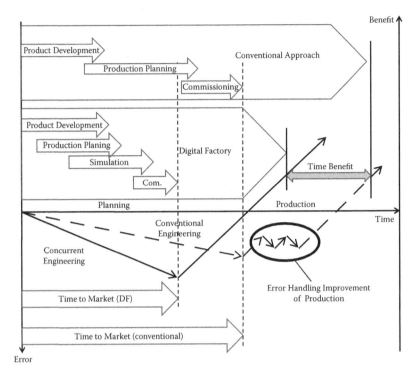

Figure 9.4 Digital Factory: benefit and effort (Adapted from Kuehn, W. (2008).)

Table 9.1 Examples of course of delays

Course of Delays: Examples	
Loss of production	**Ramp-up delay**
• Rejects due to incorrect operation of the NC machining center • Interrupted supply of utilities (electricity, gas, water, etc.) • Logistical obstacles to the supply of materials (e.g., accidents, delays at customs)	• Delays in coordinating with additional production lines • Training of machine operators scheduled at short notice • Delay in commissioning transferred machines • Sample production for customer approval

Production downtime:
  • Design changes
  • Poorly reassembled machines
  • Insufficient supply of spare parts
  • Materials held at customs
  • Late discovery of quality defects in materials supplied

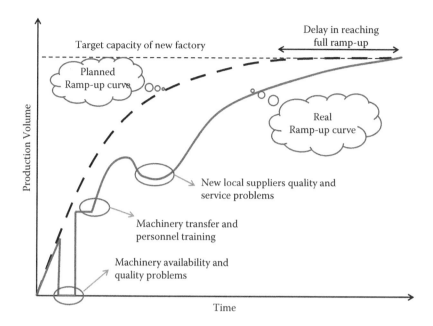

Figure 9.5 Planned and actual ramp-up curve. (Adapted from Abele, E. et al. (2008) *Global production: A handbook for strategy and implementation.* Springer, Heidelberg, Germany.)

According to the author, usually the minimum investment and maximum market impact in short time at minimum operations cost are considered in the first step of the implementation in a new location.

The framework in Figure 9.6 is useful for keeping track of the possible alternatives.

Sequence of the Ramp-Up Process

A realistic project management strategy may consist of designing a **sequential ramp-up process in order to learn from the implementation** in the first stage so to then be able to apply the improvements and eliminate the problems in subsequent steps.

Some authors (e.g., Abele et al., 2008) propose a sequential process. There are four main categories of the ramp-up process (Figure 9.7).

In ramp-up *Strategy 1*, sequential introduction of the product range **enables staff and**, if applicable, **suppliers at a new location** to successfully prepare themselves for new, complex products and their requirements. This is particularly **suitable for assembly products and production lines**. It also means that product-specific lines can be correctly industrialized (Figure 9.8).

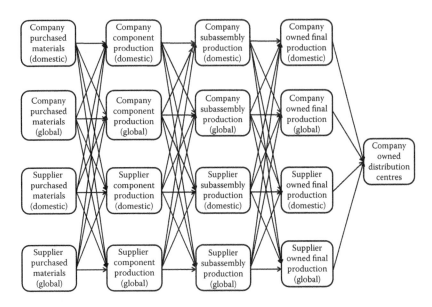

Figure 9.6 Shown are the supply chain global/local company/supplier configuration alternatives.

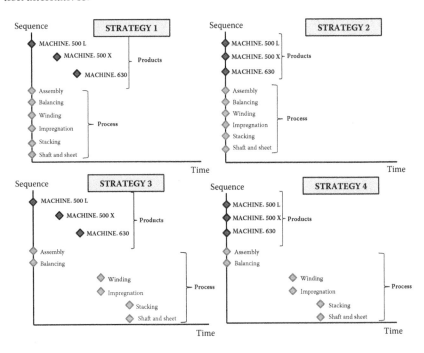

Figure 9.7 Ramp-up variants, using the example of wind power generator assembly, are illustrated.

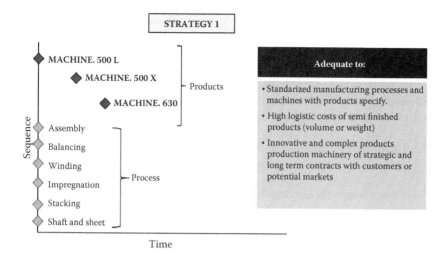

Figure 9.8 Ramp-up Strategy 1. (Adapted from Abele, E., et al. (2008) *Global production: A handbook for strategy and implementation.* Springer, Heidelberg, Germany.)

Strategy 2 involves introducing the entire range of **products and manufacturing processes simultaneous**ly. This method is suitable **only** if products and manufacturing processes are fairly **simple, or if staff is extremely skilled** and highly trained (Figure 9.9).

In *Strategy 3*, the most incremental steps possible are planned and executed. Products and manufacturing processes are introduced sequentially

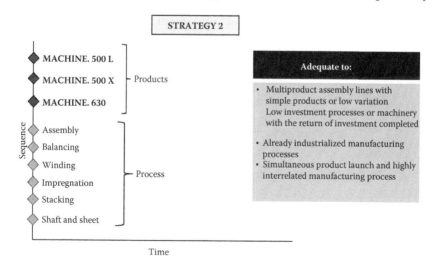

Figure 9.9 Ramp-up Strategy 2. (Adapted from Abele, E., et al. (2008) *Global production: A handbook for strategy and implementation.* Springer, Heidelberg, Germany.)

by reducing the complexity of the individual steps. Gaps in employee training can be filled successfully. This approach only makes sense if products and processes are very demanding. One disadvantage is the very **long ramp-up curve** (economies of scale are not realized until the late stage). The approach does, however, provide **high process reliability and control over the standard of quality achieved**. This third mode can be divided into two basic variants:

1. Introducing processes product by product
2. Introducing products process by process

The optimum choice depends on where the steepest learning curve or greatest economies of scale are expected (Figure 9.10).

Strategy 4 introduces products simultaneously, but production steps are introduced sequentially. This method is recommended for very **diverse, complex manufacturing processes with high quality require- ments**. It also enables the **simultaneous market launch** of a full spectrum of locally manufactured products. The start of cell phone production in new locations generally uses this method. This strategy also can be used for involving suppliers. It is advisable to introduce technically demanding primary products from new local suppliers sequentially. This allows the resolution of any technical problems to be spaced out, rather than risking multiple issues arising all at once (Figure 9.11).

Equipment Transfer Considerations

The key factors in successful equipment transfers are noted in Figure 9.12.

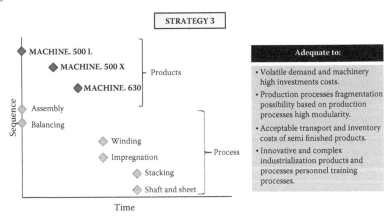

Figure 9.10 Ramp-up Strategy 3. (Adapted from Abele, E., et al. (2008) *Global pro- duction: A handbook for strategy and implementation.* Springer, Heidelberg, Germany.)

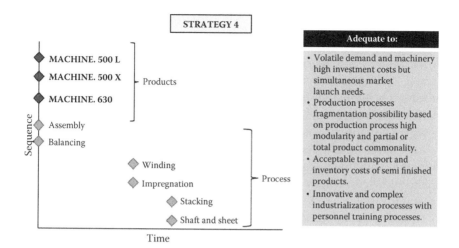

Figure 9.11 Ramp-up Strategy 4. (Adapted from Abele, E., et al. (2008) *Global production: A handbook for strategy and implementation*. Springer, Heidelberg, Germany.)

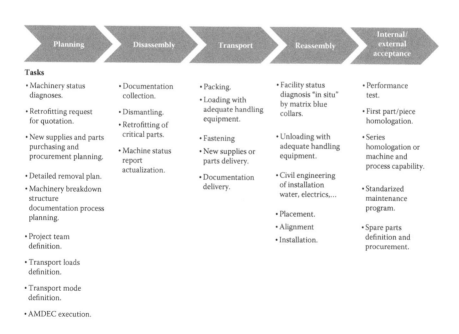

Figure 9.12 Key factors in successful equipment transfer. (Adapted from Abele, E., et al. (2008).)

Technology Adaptation Process

The use of alternative manufacturing technologies is also a key issue in the process engineering phase. Some authors (e.g., Corti, Egaña, and Errasti, 2008) state that when transferring or redesigning a manufacturing process the automation level should be revised. The engineering department should address what the reasons are for this automation, especially as regards quality or cost.

The level of automation should change if the main reason is "cost reduction in a high wage country" when transferring the production line to a low-cost country. On the other hand, if the main reason is "quality improvement" and the reason for process automation has nothing to do with cost, the process should have a similar design (Figure 9.13 to Figure 9.15).

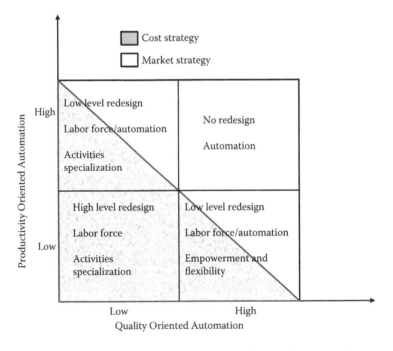

Figure 9.13 Redesign of process automation level considering quality or cost issues. (Adapted from Corti, D., et al. (2008) Challenges for off-shored operations: Findings from a comparative multi-case study analysis of Italian and Spanish companies. Paper presented at the EurOMA Congress, Groningen, The Netherlands, June 15–18.)

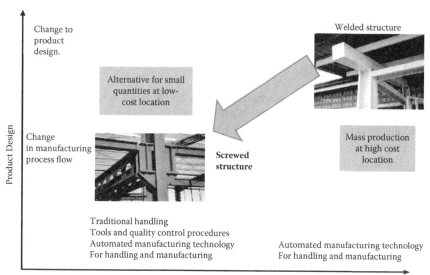

Figure 9.14 Example of alternative manufacturing methods with a change to the product design.

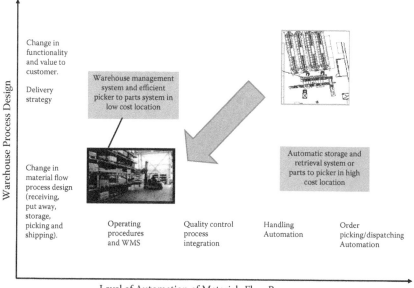

Figure 9.15 Example of alternative warehousing methods with a change to the process design.

Contingency Planning and Robust Project Management

While carrying out the ramp-up process, risk analysis and permanent monitoring of progress are critical.

During the ramp-up process, time feasibility and minimum deviation are crucial to avoid late market launch costs due to time to market or time to volume nonfulfillment. That is why companies usually prioritize time over cost, even if it means sacrificing this second variable and increasing the amount of people and extra resources needed to reach the goal (Figure 9.16).

It must be noted that world-class project management practices, such as concurrent engineering principles (Errasti et al., 2005), may be applicable, as well as time- and constraint-based project management practices, the so-called critical chain (Goldratt, 1997; Graham, 2000), even in SMEs (Apaolaza, 2009).

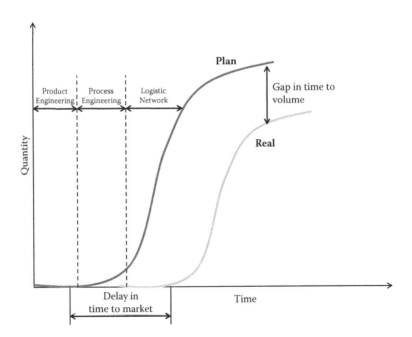

Figure 9.16 Planned and real ramp-up process and profit deviation.

References

Abele, E., Meyer, T., Näher, U., Strube, G., and Sykes, R. (2008) *Global production: A handbook for strategy and implementation*. Springer. Heidelberg, Germany

Apaolaza, U. (2009) Investigación en el método de Gestión de entornos multiproyecto "Cadena Crítica." PhD diss., Mondragon Goi Eskola Polytechnic, Spain.

Corti, D., Egaña, M. M., and Errasti, A. (2008) Challenges for off-shored operations: Findings from a comparative multi-case study analysis of Italian and Spanish companies. Paper presented at the EurOMA Congress, Groningen, The Netherlands, June 15–18.

Errasti, A., Beach, A., Oyarbide, A., and Santos, J. (2007) A process for developing partnerships with subcontractors in the construction industry. IJPM, 25 (3): 250–266.

Errasti, A., and Egaña, M. M. (2008) Internacionalización de operciones productivas: Estudio Delphi. CIL SO1, San Sebastián, Spain.

Errasti, A., Oyarbide, A., and Santos, J. (2005) Construction process reengineering. Paper presented at the proceedings of FAIM (Flexible Automation and Intelligent Manufacturing) Conference, Bilbao, Spain.

Goldratt, E. M. (1997) *Critical chain*. The North River Press, Great Barrington, MA.

Graham, K. R. (2000) Critical chain: The theory of constraints applied to project management. *International Journal of Project Management* 18: 173–177.

Kuehn, W. (2008) Digital factory — Integration of simulation enhancing the product and production process towards operative control and optimization. *International Journal of Simulation* 7(7).

Kurttila, P., Shaw, M., and Helo, P. (2010) Model factory concept: Enabler for quick manufacturing capacity ramp-up. Research paper.

Rudberg, M. and West, M. B. (2008) Global operations strategy: Coordinating manufacturing network. *Omega* 36: 91–106.

Spath, D., and Potinecke, T. (2005) Virtual product development: Digital factory based methodology for SMEs. *CIRP Journal of Manufacturing Systems* 34 (6): 539–548.

Terwiesch, C., and Bohn, R. (2001) Learning and process improvement during production ramp-up. *International Journal of Production Economics* 70 (1).

T-Systems (2010) White paper on ramp-up management. *Accomplishing full production volume in-time, in-quality, and in-cost*. Frankfurt, Germany.

VDI (Veren Deutscher Ingenieure) (Association of German Engineers Guidelines) (2006) Digital factory fundamentals. VDI 4499, Blatt 1, online at www.vdi.de.

Further Reading

Sheffi, Y. (2006) *La empresa robusta*. Ediciones Lidl, Madrid, Spain.

chapter 10

Distribution Planning and Warehousing Material Flow and Equipment Design

Claudia Chackelson and Ander Errasti

> *If you don't build your dream, someone will hire you to help build theirs.*

> **Tony Gaskins**

Contents

Introduction

In this chapter, we discuss:

- Warehouse design considerations when designing a new facility offshore
- Warehouse planning process and material flow and equipment design considering the picking volume and complexity
- Distribution models and channels
- Transport forms and route decisions

Distribution Model

Agents of the Supply Chain

Waters (2003) establishes a different Western consumer route to market models when investing in the Asian markets. These practices take into account the difficulties, cost, and time needed to develop a retail network starting from zero. Thus, the models in Table 10.1 are suggested to penetrate new markets.

Table 10.1 Distribution models when facing a distribution in a new country

Models	Trade Sales By	Comments
Own retail		Rare and not usual
Joint venture		Appropriate for strong and volume retail formats
Own logistic resources to serve retail	Brand Owner	Appropriate for dominant brands
Full agency distributor	Distributor	Appropriate for initial market entry; not always appropriate for Brand Owners
Direct export to key account retailers and rest via local distributor	Brand Owner for key accounts	
100% via local distributor	Mainly by distributor	For lower volume brands
Direct marketing	Agents	
Wholesalers	Variable	For major brands, used to serve smaller accounts via main distributor and its warehousing and transportation facility

Source: Adapted from Waters, D. (2003) *Global logistics and distribution planning: Strategies for management*. Edoiciones Kogan Page, London.

Distribution Network Configuration

There are a wide range of alternatives when designing distribution from a manufacturing facility to the customer.

The two main factors, once the supply strategy (make-to-order, assembly-to-order, make-to-stock) has been decided, are, on the one hand, the delivery volume and order complexity, and, on the other hand, service requirements in time and reliability.

Taking into account these variables and the decision of centralizing or decentralizing the stock in regional hubs near the customers, there are four different, main distributions alternatives (Abele et al., 2008) (Figure 10.1).

Introduction to Warehousing Design Considerations

Warehouses are critical nodes of a global manufacturing network from the supply chain system perspective. If we want to implement the most efficient principles, such as demand-driven supply chain (Christopher, 2005), the warehouse design should consider the management principles

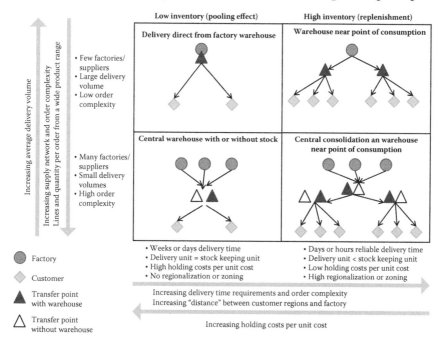

Figure 10.1 Distribution network configurations options with make-to-stock production systems. (Adapted from Abele, E., et al. (2008) *Global production: A handbook for strategy and implementation*. Springer, Heidelberg, Germany.)

oriented to increase customer satisfaction, reduce total cost, and increase return of capitals.

This chapter will follow the same structure as Chapter 6, where the general aspects of facilities are examined. The logistical activities are necessary in all kinds of production facilities. Warehouses are considered like manufacturing factories and, even if they do not manufacture, produce, or assemble, these infrastructures have great complexity and the design and management systems to be applied could be similar to manufacturing facilities when delivering orders to customers. This approach is called the *Order Factory*:

According to Errasti (2010), "The warehouse or logistic platform is an Order Factory, where once the different distribution channels and customer logistic requirements are identified, the Order Factory has to 'manufacture' and fulfill the orders promised to customers through operations and movements to enable a value-added supply process (avoiding waste), a capable process (standardized material flow and operational logistic processes), a flexible system (adaptive to demand and with a structure operational planning system), and with developed capabilities, which guarantee the availability of equipments and competent people to execute the operations."

**SUPPLY CHAIN MANAGER AND HIS/HER
SUPPORT TO THE BUSINESS STRATEGY**

The supply chain manager of the Order Factory could ask the following questions to settle a bottom-up operations strategy deployment:

- **How could the logistical strategy of the warehouse or distribution center be formulated and deployed in order to be aligned with the business strategy?**
  - New distribution channels (e.g., Internet)
  - Delivery reliability
  - Response time
  - Information visibility (e.g., stocks availability)
- **Which processes and cost drivers could be modified to increase the productivity in order to sustain a cost leadership strategy?**
  - Outsourcing of the nonprofitable activities
  - Maximize assets and resources utilization
  - Stocks reduction
  - Reduction of the damaged product

Warehouse Needs: Role or Function

A warehouse has to fulfill the customers' orders in quality and time, delivering the right goods in the right place at the right time (Rushton, Croucher, and Baker, 2006). To be able to deliver this service, warehousing (storing goods) is required as its main function.

Some of the reasons and requirements for a warehouse (Simchi-Levi, 2001; Lambert, Stock, and Ellram, 1998) in the supply network include:

- **Supply chain lead time versus delivery time gap**: The supply, manufacture, assembly, and transportation time of the supply chain generates a process lead time that could not be affordable to offer an attractive customer delivery time. Thus, a make-to-stock system from a warehouse could be a value-added service for quick response to customer needs, or operating close to them.
- **Product wider range and variety**: Customers not only search for short delivery time, but also a wide range and variety of products. Different product manufacturing and transportation times cannot be coordinated in an efficient way without a consolidation point.
- **Economies of scale**: Production and transportation economies of scale force companies to produce and move products in bigger quantities than customer needs. Thus, the warehouse couples the economic lot sizes required to produce and transport the quantities to deliver to customers.
- **Balancing supply and demand**: The seasonality and difficulty to predict the demand of some products requires the capability to deliver big quantities of products in short periods of time and to develop efficient replenishment systems to meet customers changing demands.

Related to the role or function of the warehouse, Olhager (2003) identifies four supply strategies for manufacturing companies depending on the decoupling point: engineer-to-order, make-to-order, assembly-to-order, and make-to-stock. All of these supply strategies, other than engineer-to-order, require finished goods, stock in process, and raw material warehouses. Thus, the role of the warehouse depending on the stock location could be (Figure 10.2):

- **Raw material and component warehouse**: Store and supply raw materials and components to the manufacturing facility.
- **Stock in process warehouse**: Store and supply work in process along the production and assembly process. This warehouse also could be needed when outsourcing manufacturing intermediate activities and fragmented manufacturing processes between two facilities occur.

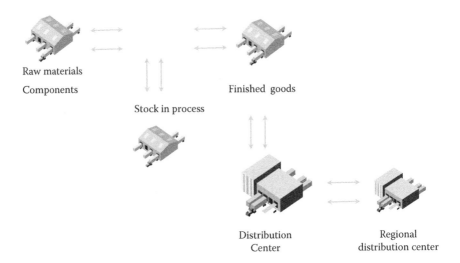

Raw materials

Components   Finished goods

Stock in process

Distribution   Regional
Center   distribution center

Figure 10.2 Role of the warehouse depending on the stock location.

- **Finished goods warehouse**: Stores finished goods that are located near the manufacturing facility.
- **Distribution center**: Stores and delivers a wide range of products to different distribution channels from stock or coordinating with manufacturing facilities in cross docking systems.
- **Regional distribution center**: Distributes within a region and enables a quick response to customer's demand.

Once the facility location (see Chapter 4) and strategic role or functions are settled, the next step is the facility design.

Tompkins (2003) and Muther (1973) state that the elements of a facility consist of the facility systems, the layout, and the handling systems. The facility systems consist of the structural system, the enclosure systems, the lighting, electrical, communication systems, etc. (Figure 10.3).

The layout consists of equipment and machinery distributed in the production and logistic areas, support areas, etc., within the building (Figure 10.4).

Warehouse Processes

The following activities are performed in all warehousing operations:

- **Receiving**: Involves accepting material, unloading, verifying quantity and condition of the material, and documenting this information as required.

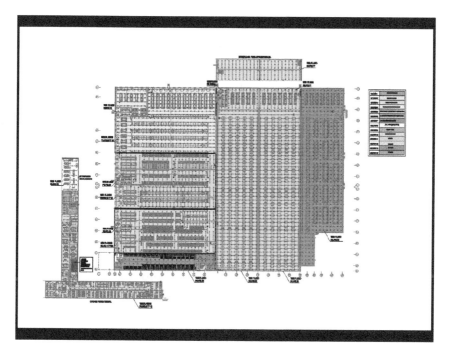

Figure 10.3 Fire water spread system in a distribution center.

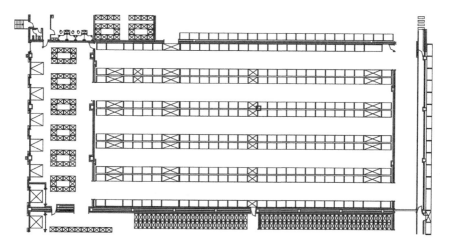

Figure 10.4 Layout of a distribution center.

- **Put away**: Means removing the goods from the receiving dock, transporting them to a storage area, locating in a specific place, and identifying where the material has been placed.
- **Storage**: The retention of products for future use or shipment.
- **Replenishment**: Occurs when material is relocated from storage to a temporary resupply from which orders are directly filled.
- **Picking and packing**: Extracting the required quantity of specific products for movement to a packing area where the goods are placed into an appropriate container, labeled with customer shipping requirements, and moving the material to a shipping area.
- **Shipping**: Involves checking quantity and condition of the material and documents and loading.

The activity that causes more operational costs and has the biggest impact on quality service is the picking process. The picking or preparation operation consists of talking and collecting articles in a specified quantity before shipment to satisfy customer orders (De Koster, 2004; De Koster et al., 2007). This is why warehouse design aspects should consider equipments, materials, technology, and personnel organization alternatives in a systemic way around the picking process (Errasti, 2010).

INDUSTRIAL SUPPLY CHAINS

ULMA Handling Systems has developed a logistics project for a company in the dairy sector, in which the automation of manufacturing systems and distribution systems have come together to offer a major logistics solution. The project consists of two installations connected by an automatic transport system: the first is built into the production process and the second is located in the Logistics Center.

The solution designed for the production process, is an installation technically classified as F.A, Factory Automation. It is composed of a comprehensive conveyor system which, combined with the use of stacker cranes, enables the "bufferization" or accumulation of pallets from packaging and palletizing systems, as well as its temporary storage in cold rooms and/or heat stoves. The critical processes also have redundant systems in order to minimize possible incidents.

The control and management system developed by ULMA controls all processes and the temperature, also integrating the control processes that guarantee product quality and complete traceability of the pallets throughout the different production stages. Therefore, the different products that pass through the system contain all the information relating to the moment it checks said information: time, temperature, production batch, etc.

The installation developed for the Logistics Center, technically classified as DA, *Distribution Automation,* is in a completely refrigerated area and receives the pallets from the Casanova installation. The connection between the two installations is fully automated, not requiring human intervention to dispatch and to proceed to loading the trucks.

The pallets from the production area are directed to the automatic warehouse in which the stacker cranes proceed to the pallet location according to the criteria established for the management system.

To ensure optimal performance and further, both in storage and order preparation, a distribution hub is provided that allows, on the one hand, to directly send pallets from production to the dispatch area, if emergencies in the order preparation require it. On the other hand, the design of the transport system in the warehouse's distribution hub is such that it allows the system to respond dynamically to a possible event that hinders the storage of a pallet at the location initially allocated. It therefore prevents a possible, although unlikely, collapse of receiving pallets from the production area.

What Is First: Layout or Handling Systems or Something Else?

Even if warehouse processes reveal the sequence of activities to be performed, the material handling solution should not be addressed after the layout macro configuration is finished and the appropriate logistics operational design is fixed.

Considering this fact and the importance of the picking process, one should consider the factors and alternatives shown in Figure 10.5.

Offshore Warehouse Design Considerations

Considering the dynamic point of view of the global markets, the new offshore static facilities design would not be suitable for the entire life cycle of the plant. This means that very few companies will be able to retain their facility or layout without severely damaging their competitive position in the marketplace. Thus, the **design should consider not only efficiency performance parameters, but warehousing reconfiguration and adaptive paradigms** (see Chapter 5).

The offshore warehouse material flow and equipment design would have to consider not only efficiency parameters, but also the ability to continually update its operations through re-layout and rearrangement. Besides, managers generally believe that a facility and management system could be replicated (or "copy-pasted") in any other site in the world.

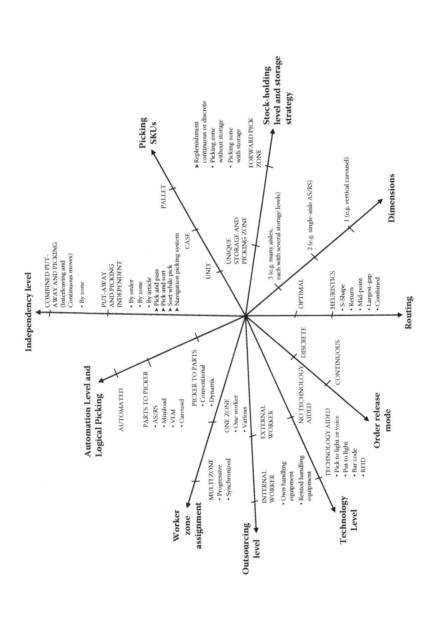

Figure 10.5 Picking systems alternatives. (Adapted from Goetschalckx and Ashayeri (1989) Classification and design of order picking systems. Logistics World, June, 99–106.).

Nevertheless, recent research (Errasti, 2009) has suggested that the process and management system modification and adaptation to local characteristics, such as labor cost, equipment maintainability, product demand variety and volumes, compared to the matrix facility, suppliers local network, etc., is a need not always taken into account.

Facilities Material Flow and Equipment Design Considerations

The same objectives of efficiency and effectiveness are applicable for warehousing (Chapter 6).

In addition, considering the layout alternatives, there are several combinations between materials, machines, or personnel alternatives to move or be fixed that determine the materials flow performance and the handling systems. Tompkins (2003) proposes material flow principles to be considered in a warehouse or logistic platform design:

- Guarantee information availability to coordinate different departments in order to **avoid extra material flow activities and administration tasks** related to receiving, putting away, replenishment, picking, and shipping.
- Eliminate or reduce material movements seeking to reduce distances.
- Improve efficiency mechanizing or automating the material handling.
- Reduce the handling needs, increasing the unit load, and palletizing units' density.

Warehouse Planning Process

In Chapter 6, we have seen that the layout and the handling system should be designed simultaneously (Tompkins, 2003), but also in operational flow characteristics (De Koster, LeDuc, and Roodbergen, 2007) and that is why while collecting the basic requirements for each production area, a battery of possible alternative block layouts should be proposed before a detailed layout is developed.

In this context, Baker and Halim (2007) propose three stages to determine the configuration that performs better (Figure 10.6).

Baker and Canessa (2009) state that the activities in Table 10.2 could be arranged in each stage.

The stages proposed by Baker and Canessa (2009) have a sequential logic in the design process; nevertheless, Tompkins et al. (2003) add that this analysis could be grouped around two main decisions: place-stock (static vision) and materials flow (dynamic vision) (Table 10.3).

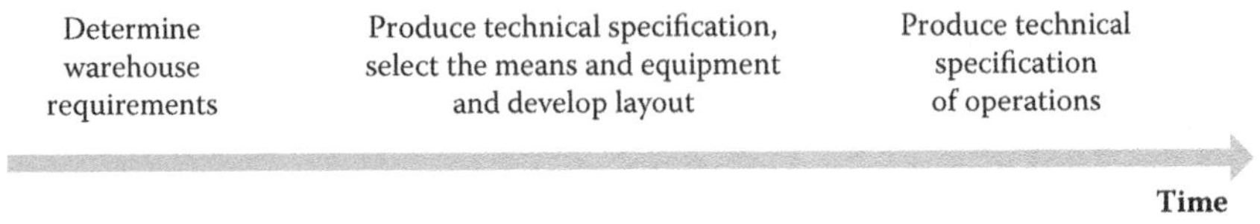

Determine warehouse requirements	Produce technical specification, select the means and equipment and develop layout	Produce technical specification of operations

Time

Figure 10.6 Stages in warehouse planning process. (Adapted from Baker, P., and Canessa, M. (2009) Warehouse design: A structured approach. *European Journal of Operational Research* 193: 425–436.)

Table 10.2 Stages and activities of the warehouse planning process

Stages	Tasks
Determine warehouse requirements	Identify the warehouse functions
	Establish unit loads to be used and form of classes (of products)
	Analyze product movement
	Determine inventory levels
	Forecast and analyze expected demand
Produce technical specification, select the means and equipment, and develop layout	Postulate operating procedures and systems
	Consider equipment types and staffing levels
	Departmentalize (into areas) and establish general layout and draft possible layout
	Calculate the space needed for storage and movement
	Calculate capital and operating costs
Produce technical specification of operations	Design storage and picking system
	Evaluate and assess expected performance
	Undertake detailed system specification/optimization
	Evaluate design against requirements

Source: Adapted from Baker, P., and Canessa, M. (2009) Warehouse design: A structured approach. *European Journal of Operational Research* 193: 425–436.

Specify the Required Material Flow Processes through Order Flow Needs

The material process "flow" shows the characteristics of the handling systems, but the logistic operational functions condition this flow. Thus, the key elements of the operational functions (demand management, service planning) that impact in the materials flow should be addressed (Figure 10.7).

More specifically, the **factors of the order picking process that impact on the complexity of the warehouse** should be identified. Usually the factors to be considered (Errasti, 2010) include:

- Type or orders (order lines, quantity per line)
- Picking volume (number of order lines, quantities, and kg-m^3)

Table 10.3 Place and flow decisions

Decision	Description
Space	Load unit and storage
	Storage equipment (racks, etc.) according to storage needs
	Handling equipment (lift truck, etc.)
	Space requirements for coordination of warehouse's processes (cross-docking, reception, put-away, storage, picking, shipping, etc.)
	Selection among different options related to productivity, quality, cost, and delivery time
Flow	Reception zone for reception flow (orders and bundle per unit time)
	Reception zone for shipping flow (orders and bundle per unit time)
	Handling equipment for reception orders versus transference according to put-away and storage zones
	Handling equipment for internal processes transference of orders (put-away, replenishment, picking, etc.)
	Handling equipment for shipping orders from sortation and accumulation zone
	Space and equipment distribution (reception, picking, shipping, etc.) according to distances minimization criteria and building limitations
	Selection among different options related to productivity, quality, cost, and delivery time

Source: Adapted from Tompkins, J. A. et al., (2010). Facilities planning, 4th ed., John Wiley & Sons, 854.

- Storage product characteristics, stock-keeping unit's dimensions: weight and variety
- Stock-keeping units versus order quantity
- Number of stored references

Flow Design Hierarchical Levels

The flow system in a warehouse is the flow of materials into the facility dock-to-dock. The flow planning hierarchy proposed by Tompkins et al. (2003) lists:

- Effective flow in workstations
- Effective flow within an area
- Effective flow between areas or layout

Flow in Workstation: Automation Just a Reason of Volume or Something Else?

One of the dilemmas when designing the manufacturing process is to determine the level of automation. It is a fact that the logistic volume

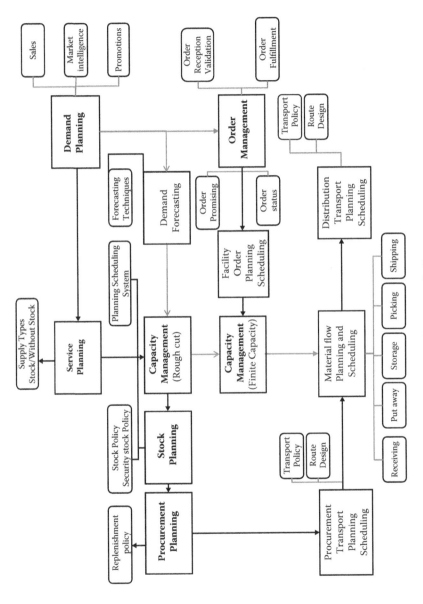

Figure 10.7 Logistic operational functions that condition the materials flow.

determines the economical feasibility and it has to do with balancing investment and operating costs in an automated, semiautomated, or manual picking system.

Workstations Flow Design with Production Volume Perspective

The picking factors related to flow activity (type of orders and picking volume) and storage (store products, stock-keeping units versus order quantity) are grouped in the following diagrams. The flow activity or volume and storage or density variables determine the appropriate solutions (rack system and picking operating methods) when considering picking systems with different levels of automation (Figure 10.8).

Workstations Flow Design Based on Quality

The order delivery quality process is also much influenced by the picking process.

In a Delphi study (Errasti and Bilbao, 2007) carried out among 42 logistic managers, the main quality failures in the order delivery process were item change, item quantity, label, damage, item extra inclusion, and order documentation (Figure 10.9).

To avoid these quality failures and increase productivity, different technologies (RF terminals, pick to light, pick to voice, etc.) and picking operating methods alternatives are introduced in medium volume density problems while integrating in a Warehouse Management System. **These technologies enable the reduction of the main quality failures**

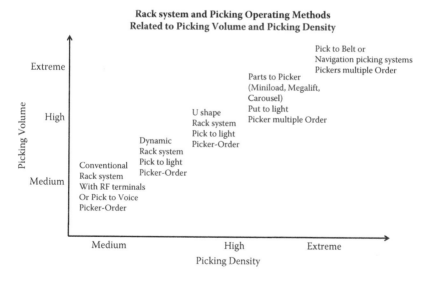

Figure 10.8 Rack systems and picking operating methods related to volume and density factors.

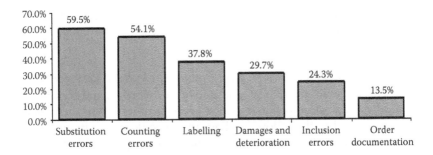

Figure 10.9 Types of order quality delivery failures. (Adapted from Errasti, A., and Bilbao, A. (2007) Proyexto OPP optimización preparación de pedidos. *Cluster de Transporte y Logística de Euskadi*, December.)

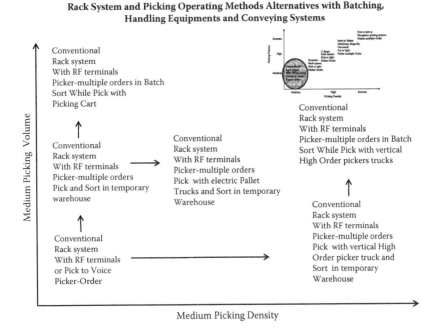

Figure 10.10 Medium volume density solutions integrating anti-error systems.

(item change, etc.) **integrating anti-error systems within the alternative picking process** (Figure 10.10) (Chackelson et al., 2012).

Workstations Flow Design Based on Variety and Volumes

When the volume and density complexity are high, there are other solutions that could be appropriate in increasing workforce productivity without automating the handling system (Figure 10.11) (Chackelson et al., 2012).

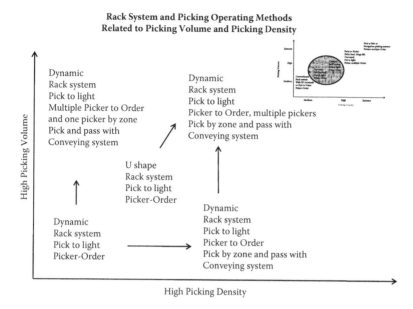

Figure 10.11 High-volume density solutions to increase workforce productivity.

Flow within an Area: Production

Areas are clusters of workstations and rack systems to be grouped together during the facility layout process. This zoning process is especially relevant in the storage area. The storage area that is the most suitable picking systems for this area and the coordination with the picking of other areas should be zoned with the factors in Figure 10.12 in mind (Frazelle, 2002).

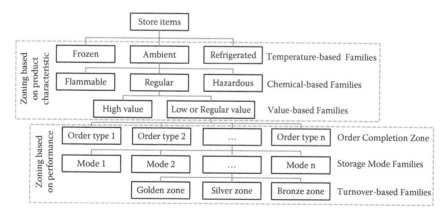

Figure 10.12 Warehouse zoning decision tree. (Adapted from Frazelle, E. (2002) *World-class warehousing and material handling*. McGraw-Hill, New York.)

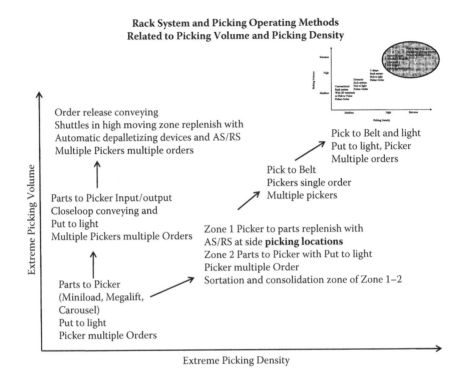

Figure 10.13 Extreme volume density solutions to increase workforce productivity.

The most relevant factors related to picking (volume and density) are customer zoning and activity or rotation.

Thus, for extreme volume–density problems, a process layout could be designed. Nonetheless, they usually need an accumulation and sorting area to consolidate the orders before shipping (Figure 10.13).

CASE ULMA Handling Systems

ALTERNATIVES AND PICKING SOLUTIONS ACCORDING TO THE COMPLEXITY OF PREPARATION

The company's order picking system is critical for its competitiveness, by always ensuring accuracy and speed of necessary services to meet client requirements.

In this regard, ULMA Handling Systems has developed crucial solutions for order picking systems in which the engineering work and technological innovation have become key elements in the development of flexible and efficient systems designed for each client.

For example, the order picking system designed for a distributor of dental products in which ULMA has designed a solution consisting of a silo with capacity to manage more than 2,000 pallets, a miniload with 17,000 containers that allow for production of 600 picking/hour carried out by two operators, a classification system and an automatic shuttle system to supply the pallet system. The company works with more than 20,000 references classified in eight product lines and through the logistic solutions implemented by ULMA, it has a capacity to serve 2,000 lines per hour.

Another example is the system developed by ULMA for a cosmetic and perfume company, in which thanks to the automation of its order picking system, the client has increased its productivity from 9,000 volumes daily to 14,000 volumes daily, reducing operational costs by 35% and achieving an increase in productivity of 60%.

Furthermore, engineering has developed a fully automated order picking system for a recognised company in the large retail sector. To give an effective solution to the company's problem with order picking, ULMA has designed a full logistics system that offers: an automatic depalletizing system; singularization, product check, orientation; automatic packing, FSS System (Fast Speed Stocker); Automatic sequencer: 600 boxes/hour; automatic depacking, sorting trays and automatic palletizing.

This is an order picking system and reference in Europe that enables a significant reduction in operating costs and the elimination of the manual process in the entire order picking system.

Transport Form or Modes

Transport connects facilities in supply chains and provides the required materials flow. The information flows have the same level of importance, which can be seen as a part of the transportation system as well. There are three basic forms of transportation: water (sea and inland water), land (road, rail, and pipeline), and air.

Which kind of transport that should be used depends on several factors, and there are often several possibilities. The one that is selected depends on the price, transportation time, service level, reliability, and additional effort, such as special packaging and handling or even additional administration time and cost.

Most of the offshore facilities must connect the demand and supply among facilities, which can have distances of thousands of miles and oceans or seas between them. In most cases, air transportation, even with the delivery time advantage, is not affordable. Thus, sea transport provides the opportunity to ship high-density products over long distances at a lower price. The disadvantages are a slow transportation speed, additional handling, and variable service levels depending on the route. **This fact increases the problems of managing long and variable lead time supply chains and the need to change Just-in-Time to Just-in-Case practices, which comes back to the need-of-stock to hide inefficiencies and problems.**

Besides the transport, incoterms (international commercial terms) have to be fixed, defining the responsibilities of the customer and the supplier along the transport process (Figure 10.14).

However, the manufacturing enterprises usually need third-party logistic enterprises to accomplish the intermodal logistic complexity (Figure 10.15).

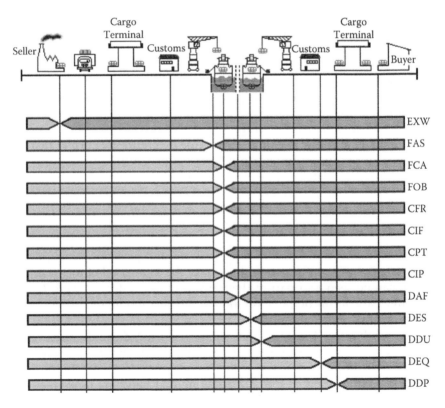

Figure 10.14 Incoterms (international commerce terms) and the customer and supplier responsibilities.

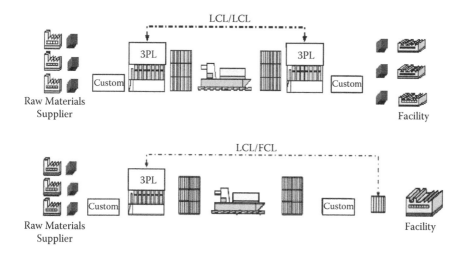

Figure 10.15 Freight management by third-party logistics in multimodal transport systems and the corresponding incoterms.

Another issue when importing or exporting goods is the selection of the harbor and the corresponding route. If we are designing a one supplier–one customer route, for example, when importing to a distribution center, the most adequate harbor is usually the nearest one. That is why the container land transport has great impact on the total transport cost and increases depending on the distance. Nevertheless, the logistic managers should request a quotation of several routes and make a purchasing effective process (Figure 10.16).

Figure 10.16 Example of a destination harbor selection depending on the distance to the distribution center.

If there are more origins and destinations, the most adequate (economic and robust) number of harbors and corresponding routes has to be designed.

In Figure 10.17, there are the options of different harbors and the corresponding routes to two distribution centers or, in Figure 10.18, one different harbor and route for each destination.

Figure 10.17 Options of different harbors and the corresponding routes to two distribution centers.

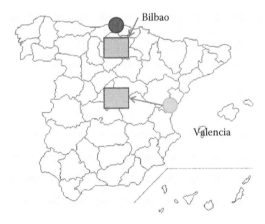

Figure 10.18 Options of one different harbor and the corresponding route for each distribution center.

INDUSTRIAL SUPPLY CHAINS

Another example designed by ULMA Handling Systems is the project developed for a textile company which integrates different processes of pre-inspection, cutting, packaging, final inspection and classification of the textile fabric which allows an effective storage system of finished, semi prepared and at dispatch products.

The process starts when the coils (fabric) are inserted into the inspection machines. Through an air and rail transport system, the coils that have gone through inspection are transported to the VTD trucks so that they are inserted into the automatic storage system.

The outputs of the coils are configured in accordance with the requirements of the client order. These coils are transported to the cutting area through the same transport system described previously. Once cut, a barcode reader detects the fate of the coil: some are packaged and others are stored for later use. Automation allows simultaneous order preparation of two trucks.

References

Abele, E., Meyer, T., Näher, U., Strube, G., and Sykes, R. (2008) *Global production: A handbook for strategy and implementation.* Springer, Heidelberg, Germany.

Baker, P., and Halim, Z. (2007) An exploration of warehouse automation implementations: Cost, service and flexibility issues. *Supply Chain Management: An International Journal* 12 (2): 129–138.

Baker, P., and Canessa, M. (2009) Warehouse design: A structured approach. *European Journal of Operational Research* 193: 425–436.

Chackelson, C., Errasti, A., and Tanco, M. (2012) A world class order picking methodology: An empirical validation. APMS 2011, IFIP AICT, 354–363. Frick J. and Laugen, B. Springer.

Chackelson, C., Errasti, A., Melacini, M., and Santos, J. (2012) Warehouse design: Validation of a new methodology through empirical research. 4th Joint World Conference on Production of Operations Management, Amsterdam, The Netherlands.

Christopher, M. (2005) *Logistics and supply chain management: Creating value-adding networks*, 3rd ed. Prentice Hall, Harlow, U.K.

De Koster, R. (2004) How to assess a warehouse operation in a single tour. Report, RSM Erasmus University, The Netherlands.

De Koster, R., Le-Duc, T., and Roodbergen, J. (2007) Design and control of warehouse order picking: A literature review. *European Journal of Operational Research* 102: 481–501.

Errasti, A. (2009) Internacionalización de Operaciones. *Cluster de Transporte y logística de Euskadi*, Diciembre.

Errasti, A. (2010) *Logística de almacenaje: Diseño y gestión de almacenes y plataformas logísticas world class warehousing.* University of Navarra, Tecnun, San Sebastian, Spain.

Errasti, A., and Bilbao, A. (2007) Proyecto OPP Optimización Preparación de Pedidos, *Cluster de Transporte y Logística de Euskadi,* Diciembre.

Frazelle, E. (2002) *World-class warehousing and material handling.* McGraw-Hill, New York.

Goetschalckx, M., and Ashayeri, J. (1989) Classification and design of order picking systems. *Logistics World,* June, 99–106.

Lambert, D. M., Stock, J. R., and Ellram, L. M. (eds.) (1998) *Fundamentals of logistics management.* McGraw-Hill, Singapore.

Muther, R. (1973) Systematic layout planning. Cahners Books, Boston.

Olhager, J. (2003) Strategic positioning of the order penetration point. *International Journal of Production Economics* 85 (3): 319–329.

Rushton, A., Croucher, P., and Baker, P. (2006) *The handbook of logistics and distribution management.* Kogan Page, London.

Simchi-Levi, D., Kaminsky, P., and Simchi-Levi, E. (2001) *Designing and managing the supply chain: Concepts, strategies and case studies.* McGraw-Hill/Irwin, Boston.

Tomkins, J. A., White, J. A., Bozer, Y., and Tranchoco, J. M. A. (2010). *Facilities planning,* 4th ed., John Wiley & Sons.

Waters, D. (2003) *Global logistics and distribution planning: Strategies for management.* Edoiciones Kogan Page, London.

Further Reading

Blackstone, J., and Cox, F. (2004) *APICS dictionary,* 11th ed. CFPIM, CIRM, Alexandria, VA.

Christopher, M., and Towill, D. (2001) An integrated model for the design of agile supply chains. *International Journal of Physical Distribution and Logistics Management* 31 (4): 23.

Errasti, A. (2006) KATAIA: Modelo para el análisis y despliegue de la estrategia logística y productiva. PhD diss., University of Navarra, Tecnun, San Sebastian, Spain.

Hoekstra, S., and Romme, J. (1992) *Integrated logistics structures: Developing customer-oriented goods flow.* McGraw-Hill, London.

Mintzberg, H. (1994) *The rise and fall of strategic planning: Reconceiving roles for planning, plans, planners.* The Free Press, New York.

Petersen, C. G., and Aase, G. (2004) A comparison of picking, storage, and routing policies in manual order picking. *International Journal of Production Economics* 92: 11–19.

Porter, M. E. (1980) *Competitive strategy: Techniques for analyzing industries and competitors.* The Free Press, New York.

Porter, M. E. (1985) *Competitive advantage.* The Free Press, New York.

Wikner, J., and Rudberg, M. (2005) Integrating production and engineering perspectives on the customer order decoupling point. *International Journal of Operations and Production Management* 25 (7): 623–640.

chapter 11

Human Resource Management

Donatella Corti

> *The only limit to our realization of tomorrow will be our doubts of today. Let us move forward with strong and active faith.*
>
> **Franklin D. Roosevelt**

Contents

Introduction

In this chapter, we discuss:

- Recruiting process
- Ramp-up process and human resource management

Recruiting Process

The success of a company competing on a global scale depends not only on the level of efficiency of operations, but also on the capability to manage, in a proper way, human resources. A multicultural workforce, in fact, needs to be carefully addressed, leading to a more complex organizational management. A survey in the *McKinsey Quarterly* (May 2008) revealed that there is a strong correlation between the ability to manage talent globally and financial performance. Even though the need for a careful management of human resources is not surprising, there are some barriers that prevent a successful HR management, such as cultural diversity, establishing globally consistent HR processes, and promoting mobility between countries (Guthridge and Komm, 2008). Findings of a more recent research on the issue of managing at global scale (Aquila, Dewhurst, and Heywood, 2012) show that over a third of interviewed respondents stated that a key element to improve operations on a global basis is to develop culturally and functionally proficient leaders.

How are companies actually managing these organizational aspects? And what are the main challenges they face? One of the difficulties could come from local recruiting. Developing countries produce far fewer graduates (white and blue collars) suitable for employment than the raw numbers might suggest.

Farrell's (2002) interviews with human resource managers for multinationals operating in low-wage countries show that only 13% of university graduates were suitable for jobs in these companies. The HR managers gave a variety of reasons for the problem including a lack of language skills, lack of practical knowledge, and a lack of cultural fit for teamwork and flexible work. This is even more relevant when it comes to offshore service operations instead of manufacturing operations. Farrell and Grant (2005) further stress this lack of local talent in China, identifying the paradox of shortages amid plenty. They state that less than 10% of job applicants are suitable for work in the analyzed service sectors in a foreign company. Main reported problems include poor English, poor communication style, low cultural fit, and lower than expected mobility of young graduates. Looking at another important country, India, the situation is slightly better. The proportion of engineering graduates who could work at a global company is as high as 25% and it is easier for companies to reach the pool of graduates in India than it is in China.

The concentration of companies in some industrial districts has some positive effects, such as developed infrastructures and communications as well as a wider availability of a workforce.

The results from a Delphi study carried out in offshore Spanish companies (Errasti and Egaña, 2009) state that in 67% of the analyzed offshore facilities the personnel rotation level is higher than in the headquarters and it depends on whether it is an emergent zone or not. The rapid growth of specific areas can create imbalances between demand and supply, and these imbalances, in turn, produce local wage inflation and turnover of workers. Furthermore, in 55% of offshore facilities, the personnel absenteeism level is higher than at the company's headquarters, in particular when in the same area there are complementary occupations, such as agriculture. These hot spots located in some cities in the Czech Republic or India should be avoided by companies in order to prevent organizational problems. The risk remains high when training local people because of high turnover, and it is also more difficult to motivate employees.

Due to problems related to the recruitment and training of local people, very often expatriates need to be brought in to plan, coordinate, control, and transfer knowledge, or to create the local team.

Current reality is that expatriates still hold a significant percentage of senior positions in the destination country, especially in positions of general management or finance. However, there is a new tendency that consists of initiating activities by expatriates and simultaneously training local managers. They then assume leadership positions and the expats are sent home to control the activity from headquarters with frequent trips back to the destination country.

A research from Baaji et al. (2012), analyzing a sample of Dutch companies, showed that the use of expatriates did not always lead to higher performance. In fact, a trade-off between strategic benefits and strategic costs needs to be carefully analyzed.

Based on the empirical findings collected by the authors, the selection process for an expatriate manager is considered a key issue, although in 40% of cases in which the expatriate moves with his family, the manager asks to be sent back home before the end of a year.

The cost for a company to send a manager to another country depends on many factors, but it is assumed that there is an increase of a 200% base salary, plus expatriation, residence, family travel, car, differential taxes, etc. (Errasti and Egaña, 2009). Another research carried out by Corti, Egaña, and Errasti (2009) related to European (mainly Italian) companies with plants in China. It was found that an expatriate manager can cost up to three times that of a local manager. In fact, expatriates keep on receiving the wage of the country of origin, get an extra bonus for moving, and, in addition, receive several benefits, such as accommodation, schools for children, and availability of an apartment in the city for weekends in case the plant is far from the city center.

The decision of having expatriated managers in China has to take into consideration not only the cost, but also other factors such as:

- Language: Local managers have to overcome the linguistic barrier.
- Cultural fit: It would be easier for local managers to get in touch with local people and, thus, achieve a good level of efficacy. Related to this point, peculiar to the Chinese market, the so-called Guanxi factor allows local managers to create a better network of relationships.
- Lack of local competences and skills.
- High turnover of Chinese managers: The shortage of qualified local managers leads to a huge increase in their wages and favors their turnover.

- Managers from headquarters in Chinese plants are often seen as a means of control where the company's strategy is interpreted correctly and the know-how is then prevented from being stolen.
- Western managers can be an investment when they are trained in supervising the setting up of offshored plants. These managers can then transfer the necessary knowledge to local managers or transfer the know-how developed locally back to headquarters.
- Fiscal and legal constraints imposed by the corporation's home country that prevents an easy move of individuals to offshored plants.

Table 11.1 summarizes the main differences of using expatriates against the use of local managers when implementing a new facility.

As mentioned by Baaij et al. (2012), there are some intermediate options a company can adopt in order to exploit the expertise of managers from the home countries while mitigating the negative aspects. Extensive usage of international solutions communication technology and travel is one possible alternative. Use of dual offices is another solution adopted by some multinational companies whose top managers have one office in the headquarters and one in the offshore plant.

"The success of a company lies in its employees." This statement, often repeated in the business world, is becoming increasingly relevant today. In an era marked by economic globalization, companies need employees more than ever to be versatile and internationally minded to execute their expansion plans abroad.

Table 11.1 Advantage of expatriate managers versus local managers

	Advantage of Expatriate Managers from Matrix	Advantage of Local Managers
General skills and qualifications	Skills and qualifications conform to standard and management techniques in the matrix	Better knowledge of local customer requirements and business practices
Product and manufacturing know-how	Knowledge of existing processes and alternatives used in the past (important in ramp-up phase) or reindustrialization of products	Local circumstances that could constraint the existing processes
Management and control	More efficient involvement in corporate and communication with central departments	
Personnel costs		Low salary level in low-ware countries; no foreign service pay or reimbursement of costs for trips home

The team in charge of the internationalization of the company should be capable of working in a hard environment characterized by uncertainty and different cultures. Therefore, a crucial skill should be the adaptability to new organizational models (new customers, new sales channels, new competition, new partners, etc.).

In this section, the results of a Delphi study (Errasti and Egaña, 2008) carried out with 15 managers will be discussed. The managers were asked about issues related to human resource management in the ramp-up process (Figure 11.1).

Results showed that, in all cases, the expatriation of white collars was necessary, and in 80% of cases there were difficulties in recruiting local managers. Recruiting of local labor was not an issue because it was considered when deciding the location. Nonetheless, the blue collars (engineers) were not aptly skilled. Thus, engineers and technicians also were expatriated.

To overcome the shortage of local talent, companies decided to reinforce the presence of expatriates in terms of both number and length of time in order to properly train the local personnel. In some cases, the travel and formation in the headquarters is an option, but its use is not widespread (Figure 11.2).

Further empirical evidence about the use of expatriates has been collected analyzing a sample of 10 Italian companies with plants in China. The number of local managers in the top management level (Figure 11.3) shows a strong correlation with the age of the manufacturing plant (Figure 11.4).

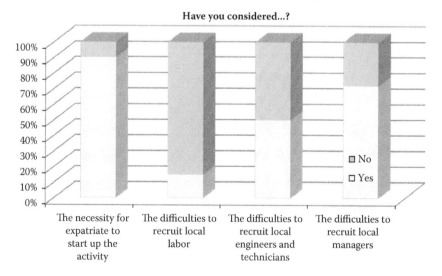

Figure 11.1 Human Resource problems in the internationalization process are shown.

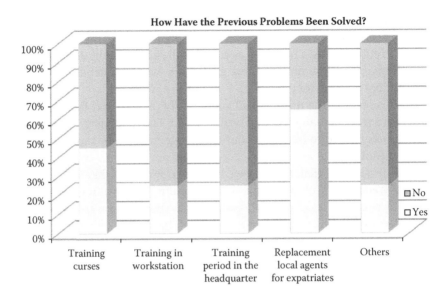

Figure 11.2 Human Resource problem solutions are listed.

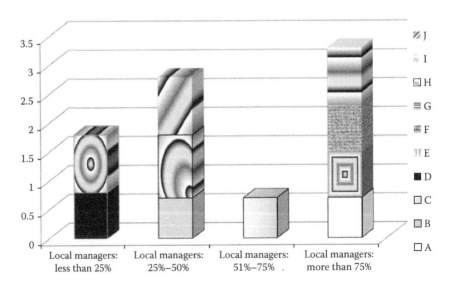

Figure 11.3 The ratio of local managers in the sampled Chinese plants.

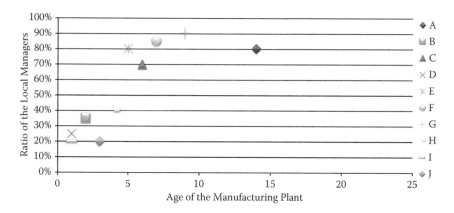

Figure 11.4 Correlation between age of manufacturing Chinese plants and ratio of local managers.

Management of Cultural Differences during the Ramp-Up Process

A research focused on the HR management and the management of cultural differences has been carried out investigating a sample of 10 Italian companies with plants in China. It appears that three main drivers have an influence on the organizational approach, namely the age of the manufacturing plant, the type of market the company competes in (B2B or B2C), and the prevalent mode of acculturation. This last driver refers to the outcome of a cooperative process where beliefs, assumptions, and values of two organizations (the headquarter and the offshored plant) form a jointly determined culture (Larsson and Lubatkin, 2001). According to a Berry study (1997), which focused on the acculturation in mergers and acquisition, the following modes have been considered:

- Integration is triggered when the members of the organizations involved in a merger want to preserve their own culture and identity and remain autonomous and independent.
- Assimilation is a unilateral process in which one group willingly adopts the identity and culture of the other.
- Separation as a mode of acculturation involves attempting to preserve one's culture and practices by remaining separate and independent from the other group.
- Deculturation occurs when the members of an organization involved in mergers do not value their own cultures and organizational practices and systems, but do not want to be assimilated into the other's practices.

Berry's model, originally proposed for mergers and acquisitions, has been adapted in this work to the setting up of a Chinese plant by an Italian firm.

The 10 case studies have been cross-analyzed and some of the empirical evidences are summarized here.

Degree of Centralization

Centralization occurs when an organization's decisions are primarily made by a small group of individuals at the top of its organization while it delegates little or no authority to the lower levels of its organization. *Decentralization* occurs when an organization distributes decision-making authority throughout its organization. A decentralized organization can be described as being loose in structure and a catalyst for open lines of communication throughout the organization, which results in an increased flow of information (Robbins and Coulter, 2003).

Seventy percent of the sampled companies adopt centralized organizational structure for the management of production plants. This shows that the power and authority are still centralized at the top management level in China. Only three companies are using a decentralized structure, but, even in this case, many production decisions are pushed down to lower levels in the organization, or even outside to some suppliers; financial and product distribution decisions still remain in the hands of senior management.

Main reasons that lead to the choice of a centralized structure include:

- The organizational structure is not mature enough to allocate the power due to the initial stage
- Quick realization of the corporate policies
- Integrity of the company

Reasons for a decentralized structure include:

- Flexibility, better adjusted to the local market
- Shorter response time to the customer
- Risk-sharing in case of the wrong decision from headquarters
- Reduction of cost and more involvement

Recruiting Process in China

Education requirements for local employees are different from expatriates. For blue collar jobs, they require a junior high school or senior high school degree, while for white collar positions, an associate's or a bachelor's degree is required.

Interesting enough is that some interviewees specified that compared with candidates who have higher education degrees, they prefer to filter them out and keep the candidates with less educational backgrounds for nonsignificant positions. The reason behind this behavior is that, the highly educated employees often take this kind of job as leverage for future career development, and the ratio of turnover is high. This could be a loss for the investment in training and resources. How to manage and maintain these employees is still an issue for the companies, and, in fact, none of the companies choose a level higher than bachelor's as the education requirement for the white collar employees' recruiting.

The length of training is similar for both blue and white collar, and it is quite short. Unlike the training in Italy, more often "days" instead of "months" is used for calculating the length of training. Therefore, long-term training is rare and only adopted for special cases.

As for the frequency of training, it has to be noted that the turnover of the blue collar workers is higher in all the sampled manufacturing plants. The training for them is basically completed at the beginning stage of admission. Frequent on-job training is considered as risky as high investment, so there is no systematic training plans for the blue collar workers. Consequently, the frequency of training for the blue collar workers is not seen as an important variable for the organizational behaviors.

Also, for the white collar training, a standardized and systematic training process is still missing for most of the companies in the sample. Apart from the initial training period, there is no systematic plan for the future training. Only two companies adopt the external sponsored training while all the other companies prefer the internal sponsored training. They still believe that the core competencies and advanced management methods are created and executed in the mother country and they did not make full use of the local resources and education. Even more and more Chinese employees benefit from the cross-sea technology and training, but the rich local resources are neglected.

As for the content of the training, it has been possible to identify four main categories for the white collar: new technology, management skills and theory, production, and manufacturing knowledge and language skills (mainly English).

Acculturation Model

The two most common managerial issues faced by the company when starting up the Chinese plants were:

- Get familiar and keep a good relationship with the local business environment, which occurs at the beginning stage of the relocation.
- The popularization and acceptance of the European management style that influences not only the initial phase, but also the entire relocation process. In fact, the companies have to take into consideration

the differences of the Chinese environmental and managerial style compared to the Western one. In some cases, the sampled companies are trying to shrink this difference by hiring the new generation.

In order to enhance the integration between the local workforce and the company's strategy, different means are used by the sampled companies.

The most adopted solution from the corporate cultural point of view is maintaining a good private relationship with the Chinese employees in addition to the job cooperation. This phenomenon is strongly supported by the Chinese culture. Because Chinese people value private relationships more than Westerners, it is significant for the employees' loyalty and satisfaction.

From the corporate policy point of view, almost half of the sample companies emphasize the importance of a competitive compensation system and wider development space for the Chinese employees. From the data of National Bureau of Statistics of China, the average salary offered by the foreign invested companies for the past 15 years is still 5.2% higher than the average salary of the related industry.

Analyzing the empirical evidences using the Berry model (1997), it has been possible to categorize companies, such as depicted in Figure 11.5. The percentage of modes of acculturation is almost equally distributed among assimilation, integration, and separation. No company uses the mode of deculturation and this can be explained considering that for the merging of two strong cultures, Italian and Chinese, it is impossible to exclude influences from the original culture to the building of a new corporate culture, and can be marked by the original features of that culture.

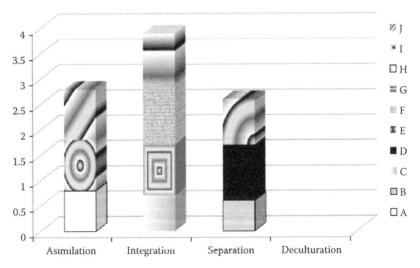

Figure 11.5 Modes of acculturation of the sampled companies.

References

Aquila, K., Dewhurst, M., and Heywood, S. (2012) Managing at global scale. *McKinsey Quarterly*, McKinsey Company, McKinsey Global Survey Results.

Baaji, M. G., Mom, T. J. M., Van Den Bosh, F. A. J., and Volberda, H. W. (2012) Should top management relocate across national borders? *MIT Sloan Management Review* winter, 53 (22): 16–19.

Berry, J. W. (1997) Immigration, acculturation and adaptation. *Applied Psychology: An International Review* 46: 5–30.

Corti, D., Egaña, M. M., and Errasti, A. (2008) Challenges for off-shore operations. Findings from a comparative multi-case study analysis of Italian and Spanish companies. EurOMA Conference.

Errasti, A., and Egaña, M. M. (2008) Internacionalización de operciones productivas: Estudio Delphi. CIL SO1, San Sebastián, Spain.

Errasti, A., and Egaña, M. M. (2009) *Research project: Internacionalización de operaciones*. Cluster de logística y transporte, San Sebastian, Spain.

Farrell, D. (2002) *Understanding the emerging global labor market*. Harvard Business School Press, Boston.

Farrell, D., and Grant, A. J. (2005) China's looming talent shortage. *McKinsey Quarterly* 4.

Guthridge, M., and Komm, A. B. (2008) Why multinationals struggle to manage talent. *McKinsey Quarterly*, May.

Larsson, R., and Lubatkin, M. (2001) Achieving acculturation in mergers and acquisitions: An international case study. *Human Relations* 54 (12): 1573–1607.

Robbins, S. P., and Coulter, M. (2003) *Management: 2003 update*, 7th ed. Prentice Hall, Upper Saddle River, NJ.

chapter 12

GlobOpe Framework

Ander Errasti

You can't do today's job with yesterday's methods and be in business tomorrow

Contents

Introduction

In this chapter, we discuss:

- Global operations (GlobOpe) design and management framework
- Improvement program once the facility is complete

GlobOpe Framework

The research on international issues in manufacturing has evolved from global business and market strategy into global operations (Johansson and Vahlne, 1977) within the value chain (engineering, manufacturing, and logistics). As competition becomes global and the complexity of the environment in which companies operate is increasing, **design and managing of the international network has become a crucial aspect** (Martinez et al., 2012).

The best network of manufacturing and supply facilities adapted to marketplaces is not an exclusive problem of multinationals operating in a worldwide context, but to small and medium enterprises (SMEs) or "late movers" that are accomplishing the internationalization process, and companies that are reconfiguring their multiple site network (Rudberg and Olhager, 2003).

Vereecke and Van Dierdonck (2002) as well as Shi (2003) state that operations and supply chain management researchers should pay attention to providing understandable models or frameworks of international manufacturing systems that help managers to design and manage their networks. Additionally, Avedo and Almeida (2011) also expose the need to build new conceptual frameworks that take into account the necessary requirements for the next generation of factories, which have to be modular, scalable, flexible, open, agile, and able to adapt, in real time, to the continuously changing market demands, technology options, and regulations.

The GlobOpe model is a framework for the design and configuration process of a global production and logistic network that can be a useful management tool for SMEs' and Strategic Business Units' (SBUs) steering committees responsible for the effectiveness and efficiency of global operations. Thus, this model aids facilities design or redesign process within a network.

The network analysis and design process is based on another approach called KATAIA (Errasti, 2006). Given the implications of configuring operations, such as new facility implementation, global supplier network development, and multiplant network reconfiguration, deciding whether and how to configure operations should be considered a strategic issue for the company, and, thus, the concepts identified in the literature were set around the steps that are typically needed in a strategy development process.

Authors, such as Acur and Biticci (2000), state that for a dynamic strategy development process four stages (inputs, analysis, strategy formulation, and strategy implementation and strategy review) are needed and that management and analytical tools can be used for this purpose.

The authors in our book have adopted this approach; nevertheless, the GlobOpe framework simplifies Acur's method and adapts it to the operational strategy business units, taking the factors stated below into consideration.

The methodology/guide takes into account the position of the business unit in the value chain (Porter, 1985) and sets the stages that should help value creation. An analysis stage is used to analyze the factors (Anumba, Siemieniuch, and Sinclair, 2000; Boddy and Macbeth, 2000; Hobbs and Andersen, 2001; Acur and Biticci, 2000) and choose the content of the strategy (Gunn, 1987). The analysis contributes to a definition or formulation of the new facility ramp-up process and then a deployment stage of the formulated design is set (Feurer, Chaharbaghi, and Wargin, 1995).

The deployment is a project-oriented task (Marucheck, Pannesi, and Anderson, 1990) where a process of monitoring and reviewing to facilitate the alignment of the organization to the operations strategy is set (Kaplan and Norton, 2001).

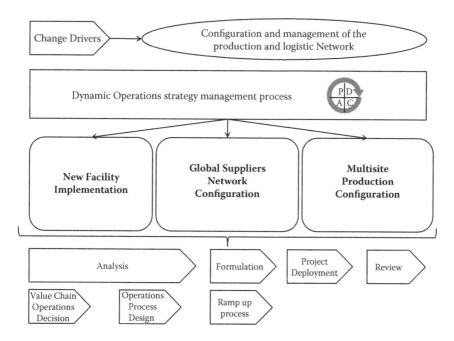

Figure 12.1 Schematic representation of the GlobOpe framework focusing on the tree problems.

As it has been explained in Chapter 2, the author of and contributors to this book consider that there are three main problems related to operations configuration and which are part of the GlobOpe Model:

- New facilities implementation
- Global supplier network configuration
- Multisite network configuration

The full framework is illustrated in Figure 12.1.

There has been developed a particular model for each one of these problems (see Chapter 2 for an explanation). These models are shown in Figure 12.2 to Figure 12.4.

The full framework is illustrated in Figure 12.5 to Figure 12.7.

GlobOpe Framework Model

This model intends to fill the gap left by the production systems (PS) (e.g., Toyota PS, Volvo PS, Bosch Siemens PS, etc.) and the Lean manufacturing programs. The Toyota PS and Lean techniques accomplish excellence in a stable environment, but they are not suitable in a market dynamic environment where new offshore facilities implementation

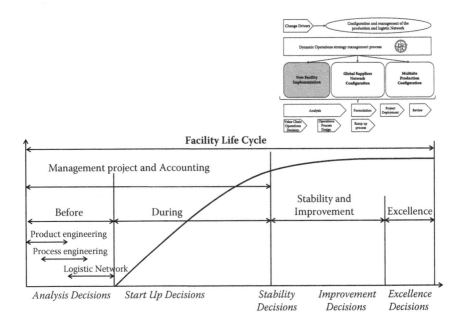

Figure 12.2 Scheme of the GlobOpe model for new facilities implementation.

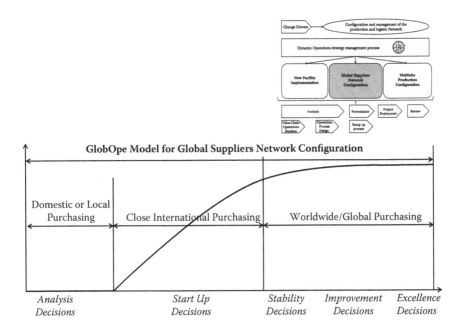

Figure 12.3 Scheme of the GlobOpe model for global suppliers network configuration.

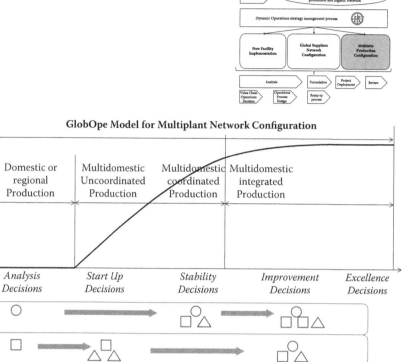

Figure 12.4 Scheme of the GlobOpe model for multiplant network configuration.

and reconfiguration of an existing network are needed (Mediavilla and Errasti, 2010). In this case, **the effectiveness and not the efficiency is the primarily goal in the first stage of implementation and more value chain activities must be taken into account when reconfiguring the network**.

This model could be a **management tool** for a Steering Committee responsible for the GlobOpe's effectiveness and efficiency.

This framework consists of new facility decision drivers, operations management principles, operations key decisions scope, and potential methods and techniques for aiding the decision process.

The principles of operations management are those paradigms that are taken into account in global operations. These are key strategic issues that managers have to decide before starting with the physical and organizational design of the operations plan, source, make, and deliver process

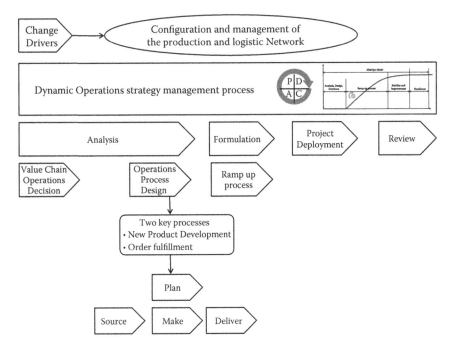

Figure 12.5 Schematic representation of the GlobOpe framework with the two key processes.

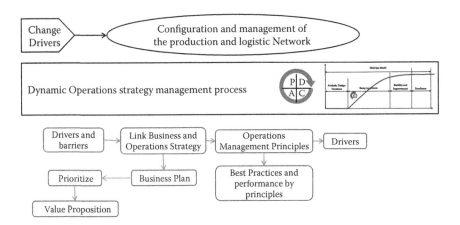

Figure 12.6 GlobOpe framework.

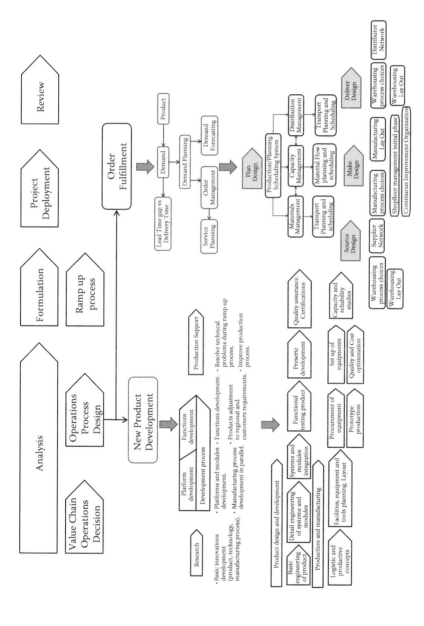

Figure 12.7 GlobOpe framework.

(Huan, Sheoran, and Wang, 2004). The methods describe the implementation process of the principles and the tools are specific aids, which are used to implement the principles and methods.

In Figure 12.8 to Figure 12.14, the GlobOpe framework is illustrated along with the chapters of the book where they are explained.

- Internationalization process business drivers, operations strategy, and business plan (Figure 12.8)
- Value chain operations decisions (Figure 12.9)
- Operations process design: Plan (Figure 12.10)
- Operations process design: Source (Figure 12.11)
- Operations process design: Make (Figure 12.12)
- Operations process design: Deliver (Figure 12.13)
- Ramp-up process (Figure 12.14)

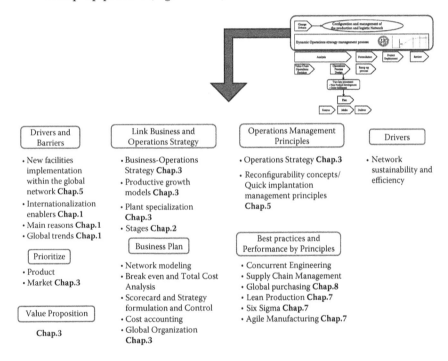

Figure 12.8 The internationalization process business drivers, operations strategy, and business plan.

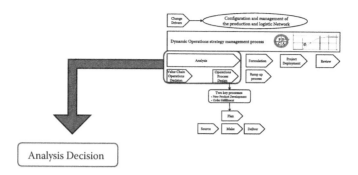

Analysis Decision

- New facility location (13 factors) **Chap.4**
- Strategic role of the plant **Chap.4**
- Productive model and type of facility **Chap.3**
- Delivery service and source strategy, MTO, MTS, ETO and main Decoupling point **Chap.3**
- Local/Global and process fragmentation (Meixal & Gargeya) **Chap.4**
- Make or buy decision to local/global network suppliers. **Chap.4**
- Productive model adaptation or management system adaptation (Yokosima, Greenfield, Brownfield) **Chap.7**
- Network design **Chap.5**

Figure 12.9 Value chain operations decisions.

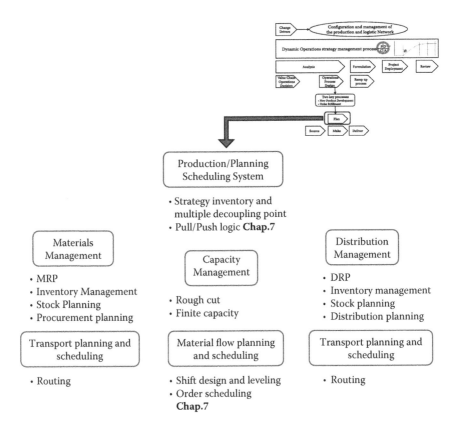

Figure 12.10 Operations process design: Plan.

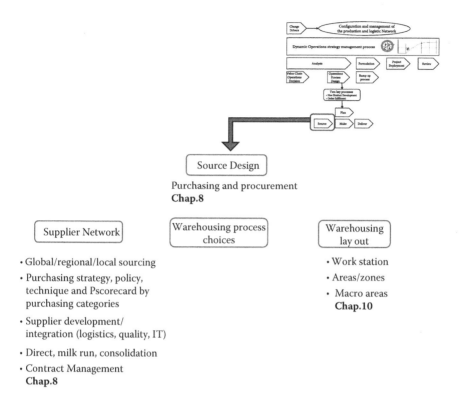

Source Design

Purchasing and procurement
Chap.8

| Supplier Network | Warehousing process choices | Warehousing lay out |

- Global/regional/local sourcing
- Purchasing strategy, policy, technique and Pscorecard by purchasing categories
- Supplier development/ integration (logistics, quality, IT)
- Direct, milk run, consolidation
- Contract Management
  Chap.8

- Work station
- Areas/zones
- Macro areas
  Chap.10

Figure 12.11 Operations process design: Source.

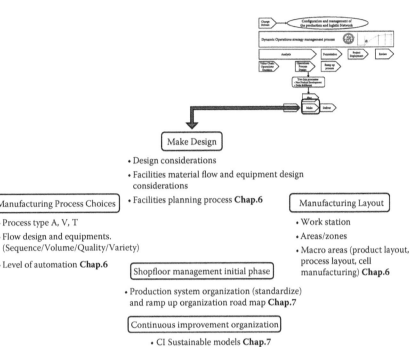

Make Design

• Design considerations
• Facilities material flow and equipment design
  considerations
• Facilities planning process **Chap.6**

Manufacturing Process Choices

• Process type A, V, T
• Flow design and equipments.
  (Sequence/Volume/Quality/Variety)
• Level of automation **Chap.6**

Manufacturing Layout

• Work station
• Areas/zones
• Macro areas (product layout,
  process layout, cell
  manufacturing) **Chap.6**

Shopfloor management initial phase

• Production system organization (standardize)
  and ramp up organization road map **Chap.7**

Continuous improvement organization

• CI Sustainable models **Chap.7**

Figure 12.12 Operations process design: Make.

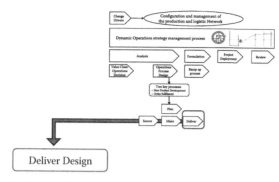

Deliver Design

- Design considerations
- Facilities material flow and equipment design considerations
- Warehouse planning process **Chap.10**

Warehousing process choices

- Process type A, V, T
- Flow design and equipments. Sequence/Volume Quality/Variety
- Level of automation **Chap.10**

Warehousing layout

- Work station
- Areas/zones
- Macro areas **Chap.10**

Distributor network

- Distribution channels
- Multilevel/echelon network and transshipment network
- Distributors and transport freight development and integration (logistics, quality, IT) **Chap.10**

Figure 12.13 Operations process design: Deliver.

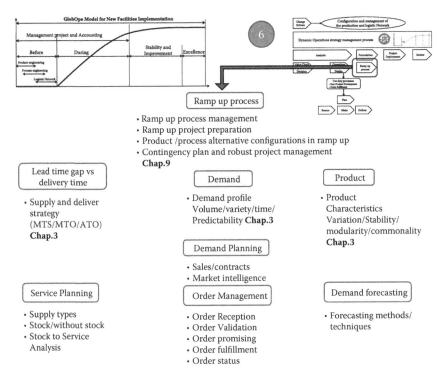

• Ramp up process management
• Ramp up project preparation
• Product /process alternative configurations in ramp up
• Contingency plan and robust project management
  Chap.9

Lead time gap vs delivery time	Demand	Product
• Supply and deliver strategy (MTS/MTO/ATO) **Chap.3**	• Demand profile Volume/variety/time/ Predictability **Chap.3**	• Product Characteristics Variation/Stability/ modularity/commonality **Chap.3**

	Demand Planning	
	• Sales/contracts • Market intelligence	

Service Planning	Order Management	Demand forecasting
• Supply types • Stock/without stock • Stock to Service Analysis	• Order Reception • Order Validation • Order promising • Order fulfillment • Order status	• Forecasting methods/ techniques

Figure 12.14 Ramp-up process.

References

Acur, N., and Biticci, U. (2000) Active assessment of strategy performance. Paper presented at the proceedings of the IFP WG 5.7 International Conference on Production Management, Tromso, Norway.

Anumba, C. J., Siemieniuch, C. E., and Sinclair, M. A. (2000) Supply chain implications of concurrent engineering. *International Journal of Physical Distribution and Logistics* 30 (7/8): 566–597.

Azvedo, A., and Almeida, A. (2011) Factory templates for digital factories framework. Robotics and Computer-Integrated Manufacturing, 27: 755–771.

Boddy, D., and Macbeth, D. (2000) Prescriptions for managing change: A survey of their effects in projects to implement collaborative working between organisations. *International Journal of Project Management* 18: 297–306.

Errasti, A. (2006) KATAIA. Modelo para el diagnostic y despliegue de la estrategia logistica y productive en PYMES y unidades de negocio de grandes empresas, PhD dissertation, TECNUN, University of Navarra, Spain.

Feurer, R., Chaharbaghi, K., and Wargin, J. (1995) Analysis of strategy formulation and implementation at Hewlett Packard. *Management Decision* 33 (10): 4–16.

Gunn, T. G. (1987) *Manufacturing for competitive advantage: Becoming a world class manufacturer.* Ballinger Publishing Company, Boston.

Hobbs, B., and Andersen, B. (2001) Different alliance relationships for project design and execution. *International Journal of Project Management* 19: 465–469.

Huan, S. H., Sheoran, S. K., and Wang, G. (2004) A review and analysis of supply chain operations reference (SCOR) model. *Supply Chain Management: An International Journal* 9 (1): 23–29.

Johansson, J., and Valhne, J. (1997) The internationalization process of the firm: A model of knowledge development and increasing foreign market commitment. *Journal of International Business Studies* 12: 305–322.

Kaplan, R. S., and Norton, D. P. (2001) *The strategy-focused organization.* Harvard Business School Press, Boston.

Martinez, S., and Errasti, A. (2012) Framework for improving the design and configuration process of an international manufacturing network. An empirical study. Frick, J. and Laugen, B. (eds.). APMS 2011, IFIP AICT, 354–363.

Marucheck, A., Pannesi, R., and Anderson, C. (1990) An exploratory study of the manufacturing strategy in practice. *Journal of Operations Management* 9 (1): 101–123.

Mediavilla, M., and Errasti, A. (2010) Framework for assessing the current strategic plant role and deploying a roadmap for its upgrading. An empirical study within a global operations network. Paper presented at the Advances in Production Management Systems (APMS) Conference, Cuomo, Italy, October 11–13.

Porter, M. E. (1985) Competitive advantage. The Free Press, New York.

Rudberg, M., and Olhager, J. (2003) Manufacturing networks and supply chains: An operating strategy perspective *Omega* 31: 29–39.

Shi, Y. (2003) Internationalisation and evolution of manufacturing systems: Classic process models, new industrial issues, and academic challenges. *Integrated Manufacturing Systems* 14: 385–396.

Vereecke, A., and Van Dierdonck, R. (2002) The strategic role of the plant: Testing Ferdows' model. *International Journal of Operations and Production Management* 22: 492–514.

Further Reading

Acur, N. (2001) Strategy management: A business process approach. PhD diss., University of Strathclyde, Glasgow.

Errasti, A., Beach, R., Odouza, C., and Apaolaza, U. (2008) Close coupling value chain functions to improve subcontractor manufacturing performance. *International Journal of Project Management* 27: 261–269.

Errasti, A., Beach, R., Oyarbide, A., and Santos, J. (2006) A process for developing partnerships with subcontractors in the construction industry: An empirical study. *International Journal of Project Management* 25: 250–256.

Index

For Product Safety Concerns and Information please contact our EU
representative GPSR@taylorandfrancis.com Taylor & Francis Verlag GmbH,
Kaufingerstraße 24, 80331 München, Germany

Printed and bound by CPI Group (UK) Ltd, Croydon, CR0 4YY
08/05/2025
01864505-0003